OFFICIAL SQA PAST PAPERS WITH ANSWERS

WITHDRAWN

HIGHER

GEOGRAPHY
2009-2013

Hodder Gibson Study Skills Advice – General — page 3
Hodder Gibson Study Skills Advice – Higher Geography — page 5
2009 EXAM — page 7
Physical and Human Environments – Environmental Interactions
2010 EXAM — page 35
Physical and Human Environments – Environmental Interactions
2011 EXAM — page 65
Physical and Human Environments – Environmental Interactions
2012 EXAM — page 97
Physical and Human Environments – Environmental Interactions
2013 EXAM — page 127
Physical and Human Environments – Environmental Interactions
ANSWER SECTION — page 155

HODDER

✕ SQA

Hodder Gibson is grateful to the copyright holders, as credited on the final page of the Question Section, for permission to use their material. Every effort has been made to trace the copyright holders and to obtain their permission for the use of copyright material. Hodder Gibson will be happy to receive information allowing us to rectify any error or omission in future editions.

Hachette UK's policy is to use papers that are natural, renewable and recyclable products and made from wood grown in sustainable forests. The logging and manufacturing processes are expected to conform to the environmental regulations of the country of origin.

Orders: please contact Bookpoint Ltd, 130 Park Drive, Abingdon, Oxon OX14 4SE. Telephone: (44) 01235 827720. Fax: (44) 01235 400454.

Lines are open 9.00–5.00, Monday to Saturday, with a 24-hour message answering service. Visit our website at www.hoddereducation.co.uk. Hodder Gibson can be contacted direct on: Tel: 0141 848 1609; Fax: 0141 889 6315; email: hoddergibson@hodder.co.uk

This collection first published in 2013 by

Hodder Gibson, an imprint of Hodder Education,

An Hachette UK Company

2a Christie Street

Paisley PA1 1NB

 Hodder Gibson is grateful to Bright Red Publishing Ltd for collaborative work in preparation of this book and all SQA Past Paper and National 5 Model Paper titles 2013.

Typeset by PDQ Digital Media Solutions Ltd, Bungay, Suffolk NR35 1BY

Printed in the UK

A catalogue record for this title is available from the British Library

ISBN 978-1-4718-0283-6

3 2 1

2014 2013

Introduction
Study Skills – what you need to know to pass exams!

Pause for thought

Many students might skip quickly through a page like this. After all, we all know how to revise. Do you really though?

Think about this:

"IF YOU ALWAYS DO WHAT YOU ALWAYS DO, YOU WILL ALWAYS GET WHAT YOU HAVE ALWAYS GOT."

Do you like the grades you get? Do you want to do better? If you get full marks in your assessment, then that's great! Change nothing! This section is just to help you get that little bit better than you already are.

There are two main parts to the advice on offer here. The first part highlights fairly obvious things but which are also very important. The second part makes suggestions about revision that you might not have thought about but which WILL help you.

Part 1

DOH! It's so obvious but …

Start revising in good time

Don't leave it until the last minute – this will make you panic.

Make a revision timetable that sets out work time AND play time.

Sleep and eat!

Obvious really, and very helpful. Avoid arguments or stressful things too – even games that wind you up. You need to be fit, awake and focused!

Know your place!

Make sure you know exactly **WHEN and WHERE** your exams are.

Know your enemy!

Make sure you know what to expect in the exam.

How is the paper structured?

How much time is there for each question?

What types of question are involved?

Which topics seem to come up time and time again?

Which topics are your strongest and which are your weakest?

Are all topics compulsory or are there choices?

Learn by DOING!

There is no substitute for past papers and practice papers – they are simply essential! Tackling this collection of papers and answers is exactly the right thing to be doing as your exams approach.

Part 2

People learn in different ways. Some like low light, some bright. Some like early morning, some like evening / night. Some prefer warm, some prefer cold. But everyone uses their BRAIN and the brain works when it is active. Passive learning – sitting gazing at notes – is the most INEFFICIENT way to learn anything. Below you will find tips and ideas for making your revision more effective and maybe even more enjoyable. What follows gets your brain active, and active learning works!

Activity 1 – Stop and review

Step 1

When you have done no more than 5 minutes of revision reading STOP!

Step 2

Write a heading in your own words which sums up the topic you have been revising.

Step 3

Write a summary of what you have revised in no more than two sentences. Don't fool yourself by saying, 'I know it but I cannot put it into words'. That just means you don't know it well enough. If you cannot write your summary, revise that section again, knowing that you must write a summary at the end of it. Many of you will have notebooks full of blue/black ink writing. Many of the pages will not be especially attractive or memorable so try to liven them up a bit with colour as you are reviewing and rewriting. **This is a great memory aid, and memory is the most important thing.**

Activity 2 — Use technology!

Why should everything be written down? Have you thought about 'mental' maps, diagrams, cartoons and colour to help you learn? And rather than write down notes, why not record your revision material?

What about having a text message revision session with friends? Keep in touch with them to find out how and what they are revising and share ideas and questions.

Why not make a video diary where you tell the camera what you are doing, what you think you have learned and what you still have to do? No one has to see or hear it but the process of having to organise your thoughts in a formal way to explain something is a very important learning practice.

Be sure to make use of electronic files. You could begin to summarise your class notes. Your typing might be slow but it will get faster and the typed notes will be easier to read than the scribbles in your class notes. Try to add different fonts and colours to make your work stand out. You can easily Google relevant pictures, cartoons and diagrams which you can copy and paste to make your work more attractive and **MEMORABLE**.

Activity 3 – This is it. Do this and you will know lots!

Step 1

In this task you must be very honest with yourself! Find the SQA syllabus for your subject (www.sqa.org.uk). Look at how it is broken down into main topics called MANDATORY knowledge. That means stuff you MUST know.

Step 2

BEFORE you do ANY revision on this topic, write a list of everything that you already know about the subject. It might be quite a long list but you only need to write it once. It shows you all the information that is already in your long-term memory so you know what parts you do not need to revise!

Step 3

Pick a chapter or section from your book or revision notes. Choose a fairly large section or a whole chapter to get the most out of this activity.

With a buddy, use Skype, Facetime, Twitter or any other communication you have, to play the game "If this is the answer, what is the question?". For example, if you are revising Geography and the answer you provide is "meander", your buddy would have to make up a question like "What is the word that describes a feature of a river where it flows slowly and bends often from side to side?".

Make up 10 "answers" based on the content of the chapter or section you are using. Give this to your buddy to solve while you solve theirs.

Step 4

Construct a wordsearch of at least 10 X 10 squares. You can make it as big as you like but keep it realistic. Work together with a group of friends. Many apps allow you to make wordsearch puzzles online. The words and phrases can go in any direction and phrases can be split. Your puzzle must only contain facts linked to the topic you are revising. Your task is to find 10 bits of information to hide in your puzzle but you must not repeat information that you used in Step 3. DO NOT show where the words are. Fill up empty squares with random letters. Remember to keep a note of where your answers are hidden but do not show your friends. When you have a complete puzzle, exchange it with a friend to solve each other's puzzle.

Step 5

Now make up 10 questions (not "answers" this time) based on the same chapter used in the previous two tasks. Again, you must find NEW information that you have not yet used. Now it's getting hard to find that new information! Again, give your questions to a friend to answer.

Step 6

As you have been doing the puzzles, your brain has been actively searching for new information. Now write a NEW LIST that contains only the new information you have discovered when doing the puzzles. Your new list is the one to look at repeatedly for short bursts over the next few days. Try to remember more and more of it without looking at it. After a few days, you should be able to add words from your second list to your first list as you increase the information in your long-term memory.

FINALLY! Be inspired...

Make a list of different revision ideas and beside each one write **THINGS I HAVE** tried, **THINGS I WILL** try and **THINGS I MIGHT** try. Don't be scared of trying something new.

And remember – "FAIL TO PREPARE AND PREPARE TO FAIL!"

Higher Geography

The course

The Higher qualification in Geography allows you to develop an understanding of the physical and human environment, and the ways in which people and the environment interact. You will develop an awareness of the changing world in a balanced, critical and sympathetic way, as well as a lifelong interest and concern for the environment. You should gain general skills in research, analysis, synthesis, data interpretation, evaluation and presentation, and develop expertise in a range of maps, diagrams, statistical techniques and information technology. Important key concepts include location, change, diversity, conflict, interdependence, co-operation, sustainability and global citizenship.

How the course is graded

Higher Geography has three Units – **Physical Environments**, **Human Environments** and **Environmental Interactions**. You are required to pass three internally marked assessments – one for each of the three Units.

The internal assessments you do in school or college (the "NABs") don't count towards the final grade, but you must have passed them before you can achieve a final grade. Your final grade will be decided in the externally marked exam in May or June.

The exam

The exam consists of two question papers. The papers are out of 100 marks each and are equally weighted.

Paper 1: Physical and Human Environments is 1 hour 30 minutes long. This paper assesses Units 1 and 2 (Physical Environments and Human Environments).

There are two sections:

- Section A consists of four **compulsory** questions. Questions 1 and 2 will be on **two** of the four **Physical Environments** topics.

  Questions 3 and 4 will be on **two** of the four **Human Environments** topics.

  These four questions will be out of between 16 and 20 marks each and one question will refer to an OS map. The topics used will vary each year.

- Section B consists of two **optional** questions. Questions 5 and 6 assess those **Physical Environments** topics not covered in Section A. You should choose **one** of these.

  Questions 7 and 8 assess those **Human Environments** topics not covered in Section A. You should choose **one** of these.

  Questions in section B are worth 14 marks each.

  Section B *may* contain a further question referring to the OS map.

Paper 2: Environmental Interactions is 1 hour 15 minutes long. It assesses Unit 3 topics. This paper consists of six questions. You must choose **two** questions based on the topics you have previously studied. Each of these is worth 50 marks.

The SQA gives detailed information for Higher Geography on its website: http://www.sqa.org.uk/sqa/38864.html

Some hints and tips

Each year, the markers are asked to draw attention to questions or topics in which students have performed particularly well, and also parts of the exam paper which could be described as common areas of weakness.

General comments

- The most obvious thing to remember is to read the question properly, and answer what you are being asked! Ensure you are familiar with the command words used and what each is asking – *explain*, for example, requires a different answer to *describe*.

- Ensure you do all that the question asks. Some questions, particularly in Paper 2, ask for a combined answer – for example '*explain management strategies to ... and discuss their effectiveness*.' Failure to answer both parts of the question will cost you marks.

- Use all of the resources given to you (maps, graphs, tables and diagrams) to help add detail to your answer.

- Know your case study material well, and ensure you can give named examples of, for example, landscape features and management strategies. If the question asks you to refer to a city/country you have studied, do so!

- Ensure you are familiar with geographical terms, such as site, situation and distribution, and be careful not to confuse words such as developed and developing, or social and environmental.

Where candidates get it right

To get the absolute most out of your Higher Geography exam, you must be sure to:

- answer all of the questions within the allocated time – practise timing yourself when completing past papers

- answer the correct choice of questions in Section B of Paper 1, and those topics you have studied in Paper 2

- write as neatly and clearly as possible in proper sentences without using simple lists or undeveloped bullet points

- draw accurate annotated diagrams where required or appropriate

- include named examples where you are asked – even when you are not, there are often marks available for being able to exemplify your answer with real life case studies

- avoid writing too much for each answer as this may cause you to run out of time later on in the paper.

Areas of success

Over the last few years, topics and questions which candidates have generally answered well include:

Paper 1:

- Physical Environments
  - explaining the formation of glacial and coastal features of erosion – candidates who are able to use annotated diagrams, in general, get more marks for these questions
  - describing and explaining plant succession along sand dunes
  - describing glacial, limestone, coastal and river landscapes using the OS map
- Human Environments
  - describing and explaining differences in rates of population growth between developed and developing countries
  - explaining reasons for migration
  - explaining changes in rural landscapes and land use

Paper 2:

- Rural Land Resources
  - as in Paper 1, describing and explaining landscape formation – those who use well annotated diagrams get more marks
  - land use conflict in upland areas
- Rural Land Degradation
  - describing processes of wind and water erosion
- River Basin Management
  - describing and accounting for the benefits and adverse consequences of a named water control project
- Development & Health
  - describing measures to manage water related diseases and the effectiveness of these strategies

Where they can get it wrong

Paper 1

Q1 Physical Environments

Some candidates find the Atmosphere topic more difficult, in particular the heat budget and redistribution of energy by ocean currents. Questions asking about global warming are often confused with the damage to the ozone layer – this is not part of the syllabus. In the Hydrosphere topic, questions asking candidates to describe and analyse hydrographs can have weaker answers. Focussing on the *processes* of erosion and deposition for all landscape formations would improve answers.

Q2 Human Environments

In the Urban topic, questions on site and situation are often poorly answered with vague responses. Many candidates find the industry questions more challenging, with some answers on industrial regeneration offering little beyond 'grants and loans'. Rural questions are usually well attempted, but make sure you answer about the farming system you have been asked about!

Paper 2

Q1 Rural Land Resources

This is one of the most popular questions in Paper 2. In general, answers on the social and economic opportunities created by the landscape could be developed beyond tourism. Remember – they must be created by the landscape!

Q2 Rural Land Degradation

If you choose this option, ensure that you are aware of both case studies. You need to be able to explain land management strategies for both North America and Africa north of the equator *or* the Amazon Basin.

Q3 River Basin Management

This is generally done well, but make sure you can describe and explain the general distribution of river basins in North America, Africa or Asia, in addition to knowing your case study well.

Q4 Urban Change and its Management

For this topic, ensure you can describe and explain distribution of urban areas. Ensure you have a sound understanding of issues related to urban sprawl, as well as those within the city in your developed world city case study.

Q6 Development and Health

This is another popular option. Ensure you read the questions properly. The development section may refer to differences in development *between* countries or *within* a country. This is confused by some.

Good luck!

Remember that the rewards for passing Higher Geography are well worth it! Your pass will help you get the future you want for yourself. In the exam, be confident in your own ability. If you are not sure how to answer a question, trust your instincts and write what you can. It's always worth a try. GOOD LUCK!

[BLANK PAGE]

X208/301

| NATIONAL
QUALIFICATIONS
2009 | WEDNESDAY, 27 MAY
9.00 AM – 10.30 AM | GEOGRAPHY
HIGHER
Paper 1
Physical and
Human Environments |

Six questions should be attempted, namely:

all four questions in **Section A** (Questions 1, 2, 3 and 4);

one question from **Section B** (Question 5 **or** Question 6);

one question from **Section C** (Question 7 **or** Question 8).

Write the numbers of the **six** questions you have attempted in the marks grid on the back cover of your answer booklet.

The value attached to each question is shown in the margin.

Credit will be given for appropriate maps and diagrams, and for reference to named examples.

Questions should be answered in sentences.

Note   The reference maps and diagrams in this paper have been printed in black only: no other colours have been used.

1:50 000 Scale
Landranger Series

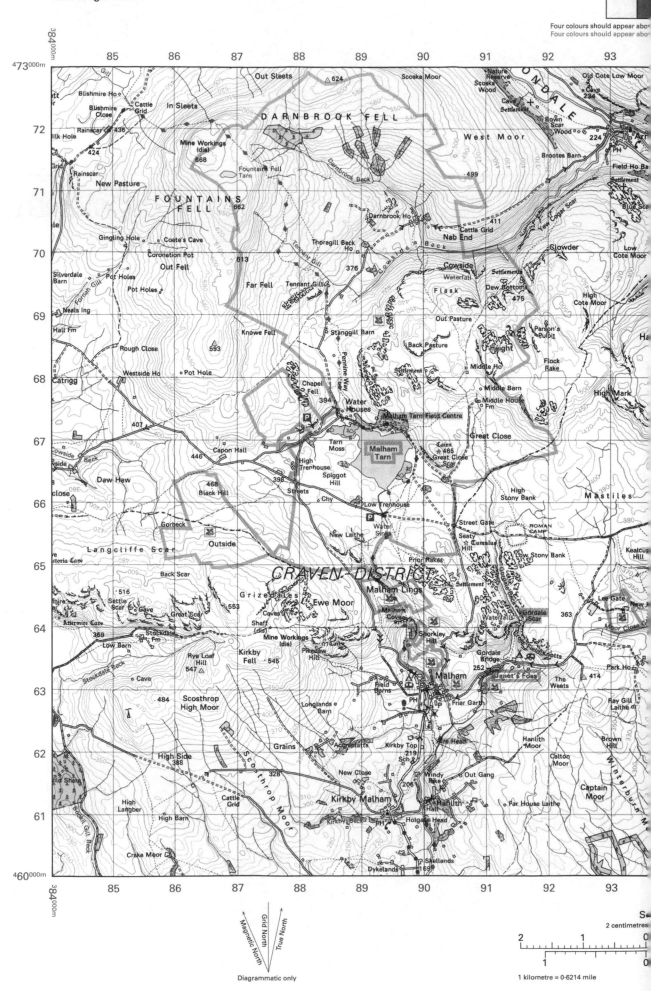

please return to the invigilator.
please return to the invigilator.

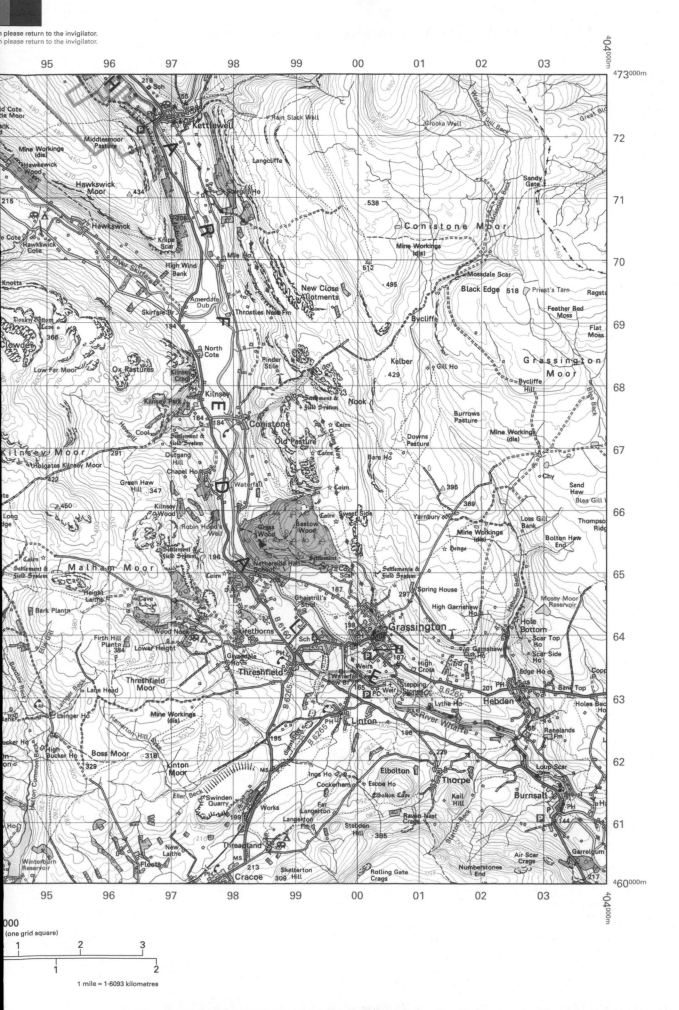

000
(one grid square)

1 mile = 1·6093 kilometres

Marks

SECTION A: Answer ALL questions in this section

Question 1: Hydrosphere

Study OS Map Extract number 1745/98: Upper Wharfedale (*separate item*).

(*a*) Using appropriate grid references, **describe** the physical characteristics of the River Wharfe and its valley from 978690 to 040603.

10

(*b*) **Explain**, with the aid of a diagram or diagrams, how a waterfall is formed in the upper course of a river valley.

8

Marks

Question 2: Biosphere

(*a*) **Draw** and **fully annotate** a soil profile of a **podzol** to show its main characteristics (including horizons, colour, texture and drainage) and associated vegetation.

9

Study Reference Diagram Q2 which shows a soil profile of a brown earth soil.

(*b*) **Describe** and **explain** the formation and characteristics of a **brown earth soil**.

9

Reference Diagram Q2

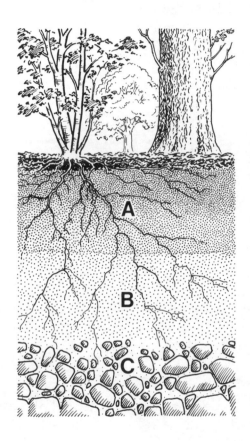

[Turn over

Marks

Question 3: Rural Geography

(a) Study Reference Table Q3.

Describe and **explain**, with the aid of the data in the table, the differences between intensive peasant farming and commercial arable farming.   **10**

Reference Table Q3 (Types of farming and selected data)

	Bangladesh	**Canada**
Farm type	Intensive peasant farming	Commercial arable farming
GDP per capita ($US)	158	13 034
% GDP from farming	48	3
% population engaged in farming	82	4
Kg of fertiliser used per hectare	26	32
People per tractor	20 581	38

(b) Areas of intensive peasant farming such as those in Bangladesh have undergone changes in recent years.

Referring to an area you have studied:

(i) **describe** these changes, and

(ii) **outline** the impact of these changes on the people **and** the farming landscape.   **10**

Marks

Question 4: Industrial Geography

"*Many industrial concentrations within the European Union have undergone a great transformation in the last 50 years. These changes are most marked in the types of industries, the industrial landscape and in employment patterns.*"

Referring to a **named** industrial concentration in the European Union that you have studied:

(i) **describe** and **account for** the main characteristics of a typical "new" industrial landscape;   9

(ii) **suggest** ways in which the national government **and** the European Union have helped to attract new industries to your chosen area.   7

[Turn over

Marks

SECTION B: Answer ONE question from this section, ie either Question 5 or Question 6.

Question 5: Atmosphere

"Energy is transferred from areas of surplus, between 35°N and 35°S, to areas of deficit, polewards from 35°N and 35°S, by both oceanic and atmospheric circulation."

Study Reference Map Q5 which shows selected ocean currents in the North Atlantic Ocean.

(a) (i) **Describe** the pattern of ocean currents in the North Atlantic Ocean, and

   (ii) **explain** how they help to maintain the global energy balance. **6**

Reference Map Q5 (Selected ocean currents in the North Atlantic Ocean)

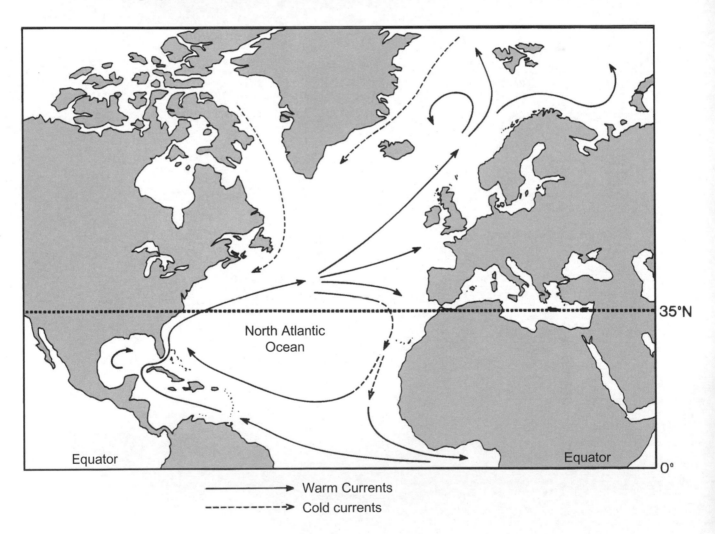

Warm Currents
--------→ Cold currents

Marks

Question 5: Atmosphere (continued)

Study Reference Diagram Q5 which shows surface winds and pressure zones.

(b) **Explain** how circulation cells in the atmosphere and the associated surface winds assist in the transfer of energy between areas of surplus and deficit.

8

Reference Diagram Q5 (Surface winds and pressure zones)

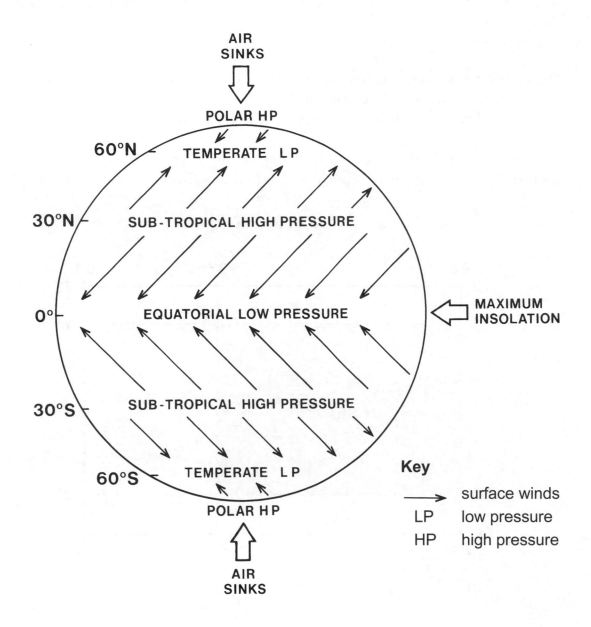

[Turn over

Marks

DO NOT ANSWER THIS QUESTION IF YOU HAVE
ALREADY ANSWERED QUESTION 5

Question 6: Lithosphere

Study OS Map Extract number 1745/98: Upper Wharfedale (*separate item*), and Reference Map Q6.

The map extract covers part of the Yorkshire Dales National Park, an area famed for its Carboniferous Limestone scenery, characterised by distinctive surface features, drainage patterns and underground landforms.

(a) **Describe** the evidence which suggests that Area A, shown on Reference Map Q6, is a Carboniferous Limestone landscape.
(You should refer to named features and make use of grid references.)   8

(b) Choose any **one** Carboniferous Limestone feature described in your answer to part (a) and, with the aid of annotated diagrams, **explain** how it was formed.   6

Reference Map Q6

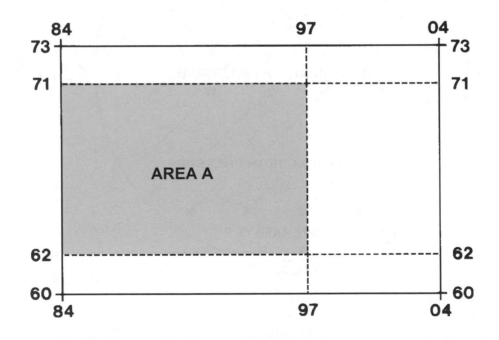

Marks

**SECTION C: Answer ONE question from this section,
ie either Question 7 or Question 8.**

Question 7: Population

Italy has a population structure that is typical of many EMDCs (Economically More Developed Countries).

Study Reference Diagrams Q7A and Q7B.

(a) **Describe** and **account** for the changes between the population structure in 2000 and that projected for 2050.

8

(b) **Discuss** the consequences of the 2050 population structure for the future economy of the country and the welfare of its citizens.

6

Reference Diagram Q7A (Italy: Population pyramid for 2000)

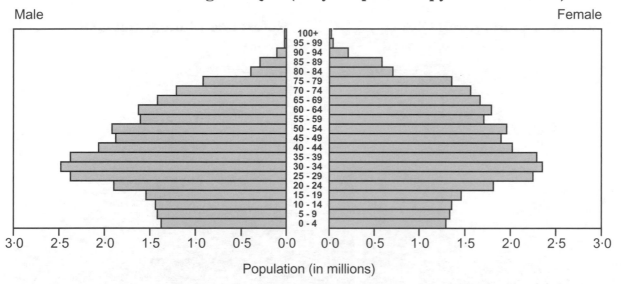

Reference Diagram Q7B (Italy: Population pyramid for 2050)

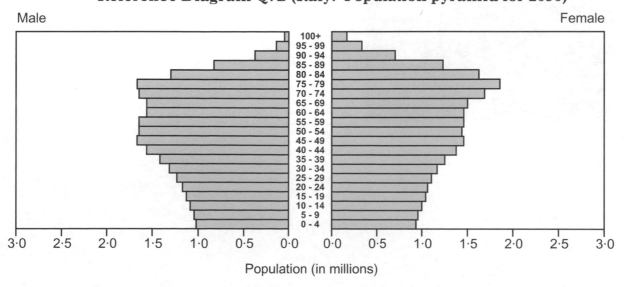

Marks

DO NOT ANSWER THIS QUESTION IF YOU HAVE ALREADY ANSWERED QUESTION 7

Question 8: Urban Geography

Study Reference Photograph Q8A which shows Buchanan Galleries shopping centre in Glasgow's CBD and Reference Photograph Q8B which shows Braehead, an out-of-town shopping centre situated at the south-west edge of Glasgow.

Referring to Glasgow, **or** any other named city you have studied in an Economically More Developed Country (EMDC):

(i)   **suggest** the impact that an out-of-town shopping centre may have had on shopping in the traditional CBD;   **6**

(ii) **describe** and **explain** the changes, other than shopping, which have taken place in the CBD over the past few decades.   **8**

Reference Photograph Q8A   **Reference Photograph Q8B**

[END OF QUESTION PAPER]

X208/303

NATIONAL
QUALIFICATIONS
2009

WEDNESDAY, 27 MAY
10.50 AM – 12.05 PM

GEOGRAPHY
HIGHER
Paper 2
Environmental
Interactions

Answer any **two** questions.

Write the numbers of the **two** questions you have attempted in the marks grid on the back cover of your answer booklet.

The value attached to each question is shown in the margin.

Credit will be given for appropriate maps and diagrams, and for reference to named examples.

Questions should be answered in sentences.

Note The reference maps and diagrams in this paper have been printed in black only: no other colours have been used.

Marks

Question 1 (Rural Land Resources)

Loch Lomond and the Trossachs became Scotland's first National Park in 2002. It covers 1865 square kilometres of lowland, river, loch, forest and mountain landscapes.

(*a*) **Describe** and **explain**, with the aid of annotated diagrams, the formation of the main glacial features of the Loch Lomond and the Trossachs National Park **or** any other glaciated upland area in the UK that you have studied.

20

(*b*) With reference to Loch Lomond and the Trossachs **or** any other named upland area that you have studied, **explain** the social and economic opportunities created by the landscape.

10

(*c*) Study Reference Diagram Q1.

Reference Diagram Q1 shows the Loch Lomond and the Trossachs National Park to be under intense environmental pressure in certain key areas. With reference to this area or any **named** upland area you have studied:

(i) **describe** and **explain** the environmental conflicts that may occur (you should refer to named locations within your chosen upland landscape);

10

(ii) **describe** specific solutions to these environmental conflicts commenting on their effectiveness.

10

(50)

Question 1 — continued

Reference Diagram Q1 (Loch Lomond and the Trossachs: Environmental Activity and Pressure)

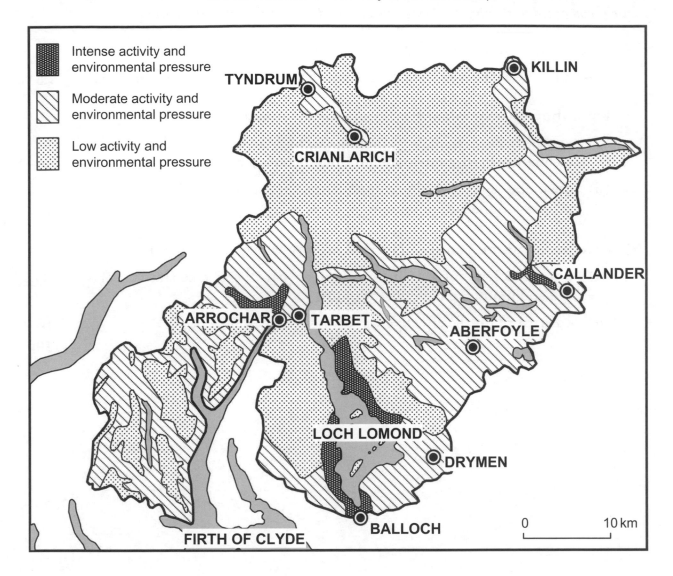

[Turn over

Marks

Question 2 (Rural Land Degradation)

The Sahel is a 500 kilometre wide zone which runs across Africa along the southern edge of the Sahara Desert. The Sahel is under intense pressure from human activity which, combined with climate change, has created a "spiral of desertification".

(*a*) Study Reference Diagram Q2.

Describe the changes in rainfall patterns shown on Reference Diagram Q2.   **6**

(*b*) For **either** Africa north of the Equator **or** the Amazon Basin:

(i) **explain** how human activities, including inappropriate farming techniques, have contributed to land degradation; and   **18**

(ii) **describe** some of the consequences of land degradation on the people and their environment.   **10**

(*c*) Referring to **named** areas of **North America** which you have studied:

(i) **describe** some of the measures which have been taken to conserve soil and limit land degradation; and

(ii) **comment on** the effectiveness of these measures.   **16**

(50)

Question 2 — continued

Reference Diagram Q2 (Rainfall Variability in the Sahel)

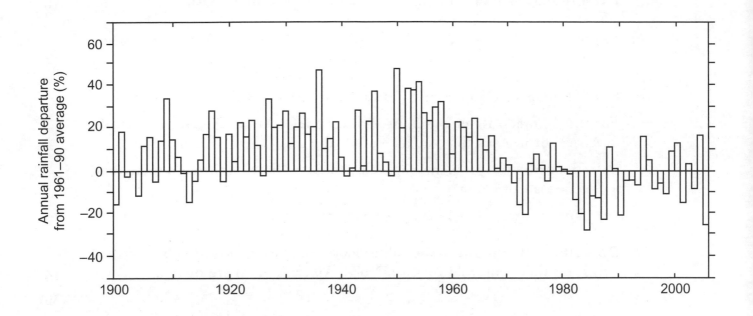

[Turn over

Marks

Question 3 (River Basin Management)

(a) Study Reference Table Q3 and Reference Map Q3.

Explain the need for water management in the Colorado Basin.   **10**

(b) **Explain** the physical **and** human factors that have to be considered when selecting sites for dams and their associated reservoirs.   **14**

(c) Study Reference Diagram Q3 and Reference Map Q3.

For the Colorado River Basin, **or** another river basin in North America, **or** in Africa, **or** in Asia, that you have studied:

(i) **describe** the problems caused by the river flowing through more than one state or country;

(ii) **suggest** ways in which these problems may be overcome.   **10**

(d) **Describe** and **explain** the social, economic and environmental **benefits** of a **named** water control project in North America **or** Africa **or** Asia.   **16**

(50)

Reference Table Q3 (Population Growth in Las Vegas and Phoenix)

Selected city	1990 Population	2000 Population	Population change (1990–2000)
Phoenix	2 238 480	3 251 876	+45%
Las Vegas	741 459	1 375 765	+85%

Reference Diagram Q3 (The Colorado River Water Allocation)

Upper Basin–Water allocation   **Lower Basin–Water allocation**

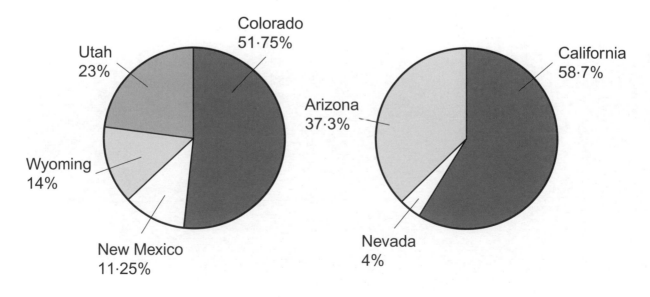

Question 3 — continued

Reference Map Q3 (The Colorado River Basin)

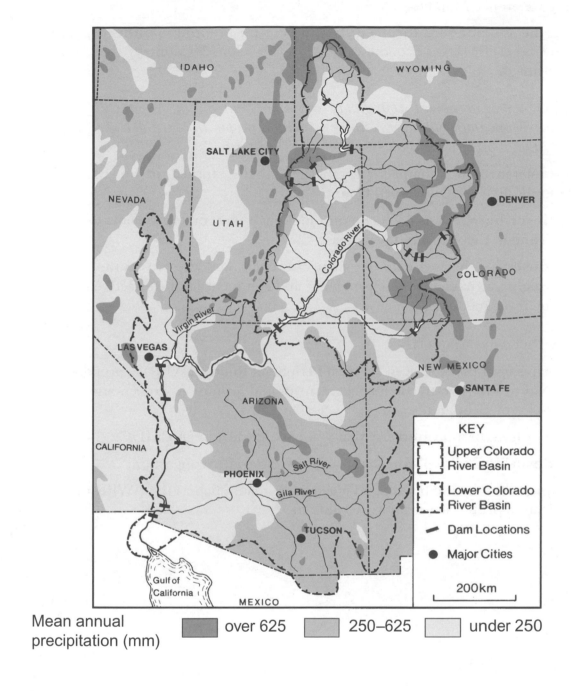

Mean annual precipitation (mm)   over 625   250–625   under 250

[Turn over

Marks

Question 4 (Urban Change and its Management)

(*a*) Study Reference Map Q4A.

Describe and **account for** the distribution of major cities in **either** Spain **or** any other EMDC (Economically More Developed Country) that you have studied.

10

(*b*) "*Kibera is one of almost 100 shanty towns in Nairobi, the capital city of Kenya. More than half of Nairobi's 3 million people live in these shanties, which in total occupy less than 2% of the city's land area.*"

With reference to a named city that you have studied in an ELDC (Economically Less Developed Country):

(i) **describe** the social, economic and environmental problems often found in these shanty town areas;

12

(ii) **describe** the methods the shanty dwellers and the city authorities might use to tackle these problems, and comment on the effectiveness of these methods.

8

(*c*) Study Reference Map Q4B.

The map shows the Aberdeen Western Peripheral Route (AWPR), a proposed new road to improve traffic management in and around Aberdeen and the North-east of Scotland.

For Aberdeen, or a **named** city that you have studied in an EMDC:

(i) **describe** and **explain** why it suffers from traffic congestion;

12

(ii) **suggest** why the building of major new roads such as the AWPR may lead to protests and land-use conflicts.

8

(50)

Question 4 — continued

Reference Map Q4A (Largest Cities in Spain)

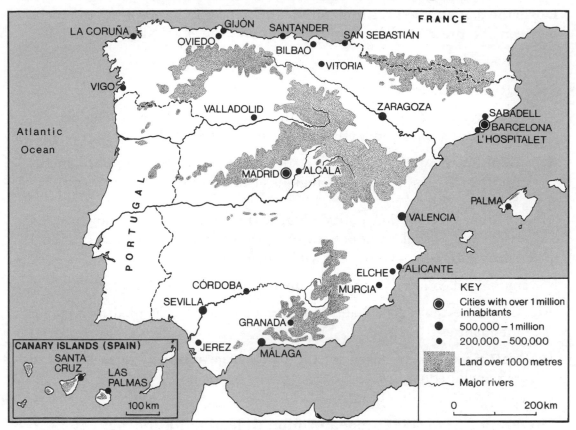

Reference Map Q4B (Aberdeen Western Peripheral Route (AWPR))

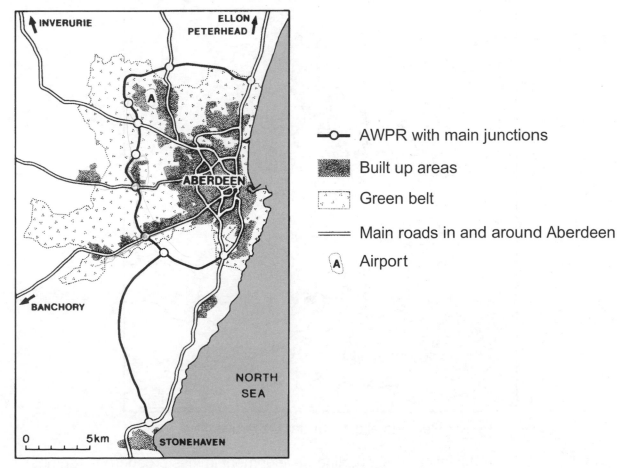

Question 5 (European Regional Inequalities)

(*a*)　Study Reference Table Q5.

　　Describe and **suggest reasons** for the differences in levels of development between the pre-2000 EU member states and the post-2000 EU member states.　　**10**

(*b*)　Study Reference Map Q5.

　　(i)　**Describe** the distribution of the regions which were eligible for European grants under Objective 1 support (2000–2006).　　**8**

　　(ii)　**Explain** how EU initiatives such as Objective 1 support might improve the less prosperous regions of the European Union.　　**8**

(*c*)　"The European Cohesion policy (2007–2013) aims to contribute towards economic and social cohesion within the EU by reducing regional differences and human inequality within member states."

　　For any named country you have studied in the European Union:

　　(i)　**describe** the physical and human factors which have led to regional inequalities;　　**18**

　　(ii)　**outline** the steps taken by the national government to tackle these regional inequalities.　　**6**

　　　　　　　　　　　　　　　　　　　　　　　　　　　(50)

Reference Map Q5 (European Union Objective 1 Funding)

　　▨　Regions eligible under Objective 1

(Objective 1:　Supporting development in less prosperous regions)

Question 5 — continued

Reference Table Q5 (European Union Statistics Ranked in Order)

Pre–2000 Member States				Post–2000 Member States			
Country	Year of EU membership	GDP (ranked)*	HDI (ranked)*	Country	Year of EU membership	GDP (ranked)*	HDI (ranked)*
Belgium	1957	6	6	Cyprus	2004	14	17
France	1957	11	9	Czech Rep	2004	17	18
Germany	1957	10	13	Estonia	2004	20	22
Italy	1957	12	10	Hungary	2004	21	20
Luxembourg	1957	1	5	Latvia	2004	24	25
Netherlands	1957	3	3	Lithuania	2004	23	23
Denmark	1973	5	8	Malta	2004	18	19
Ireland	1973	2	1	Poland	2004	25	21
UK	1973	8	11	Slovakia	2004	22	24
Greece	1981	15	14	Slovenia	2004	16	15
Portugal	1986	19	16	Bulgaria	2007	26	26
Spain	1986	13	12	Romania	2007	27	27
Finland	1995	9	4				
Sweden	1995	7	2				
Austria	1995	4	7				

GDP   Gross Domestic Product per capita reflects total of all goods and services per head of population

HDI   Human Development Index (covering poverty, education, health)

*Ranking   1–27 with 1 best and 27 worst

[Turn over

Marks

Question 6 (Development and Health)

(a) Study Reference Map Q6 which shows the Human Development Index (HDI) for countries of the world.

Explain the advantages of using a composite indicator of development such as the HDI rather than a single indicator.

4

(b) Referring to named examples, **suggest reasons** why there is such a wide range in levels of development **between** different ELDCs (Economically Less Developed Countries).

12

(c) For malaria, **or** bilharzia, **or** cholera:

(i) **describe** the human and environmental factors that can contribute to the spread of the disease;

6

(ii) **describe** the measures that have been taken to combat the disease;

12

(iii) **explain** how the eradication or control of the disease would benefit ELDCs.

6

(d) *"Resources need to be targeted at improving Primary Health Care if we are ever going to improve the health of people in ELDCs."* Aid worker

Describe some of the strategies involved in Primary Health Care and **explain** why these strategies for improving health standards are suited to people living in ELDCs.

10

(50)

Question 6 — continued

Reference Map Q6 (The World: Human Development Index)

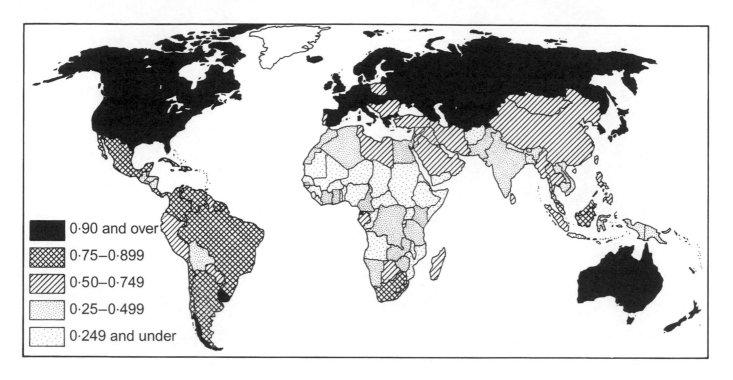

The Human Development Index measures development by combining three individual measures. These measures are:

- adult literacy rate;

- life expectancy;

- real Gross Domestic Product (ie what an income will actually buy in a country).

[END OF QUESTION PAPER]

[BLANK PAGE]

HIGHER

2010

[BLANK PAGE]

X208/301

NATIONAL QUALIFICATIONS 2010	MONDAY, 31 MAY 9.00 AM – 10.30 AM	GEOGRAPHY HIGHER Paper 1 Physical and Human Environments

Six questions should be attempted, namely:

all four questions in **Section A** (Questions 1, 2, 3 and 4);

one question from **Section B** (Question 5 **or** Question 6);

one question from **Section C** (Question 7 **or** Question 8).

Write the numbers of the **six** questions you have attempted in the marks grid on the back cover of your answer booklet.

The value attached to each question is shown in the margin.

Credit will be given for appropriate maps and diagrams, and for reference to named examples.

Questions should be answered in sentences.

Note   The reference maps and diagrams in this paper have been printed in black only: no other colours have been used.

1:50 000 Scale
Landranger Series

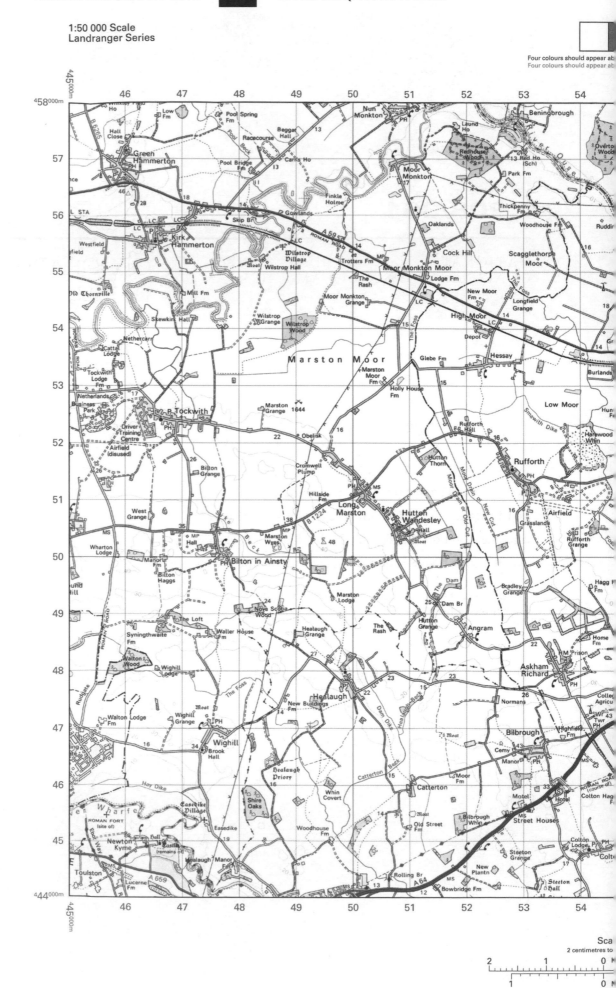

1 kilometre = 0·6214 mile

Extract No 1788/105

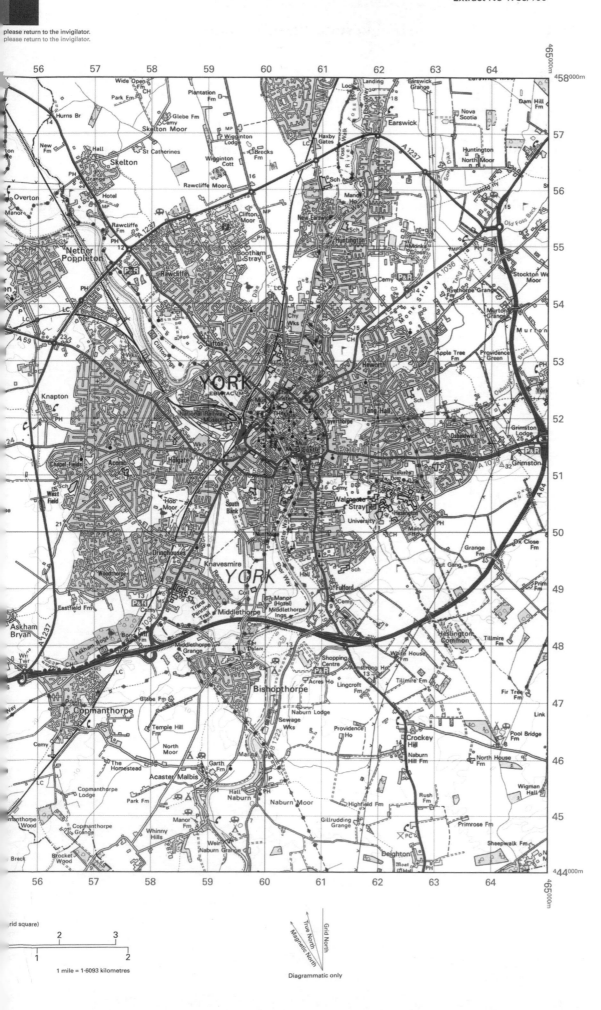

1 mile = 1·6093 kilometres

True North
Grid North
Magnetic North

Diagrammatic only

Marks

SECTION A: Answer ALL questions in this section

Question 1: Atmosphere

Study Diagram Q1.

(*a*) **Describe** the **human** factors that may lead to the global temperature projection shown in the diagram.   **10**

(*b*) **Describe** and **explain** the possible consequences of global warming.   **10**

Diagram Q1: Global Warming Projection

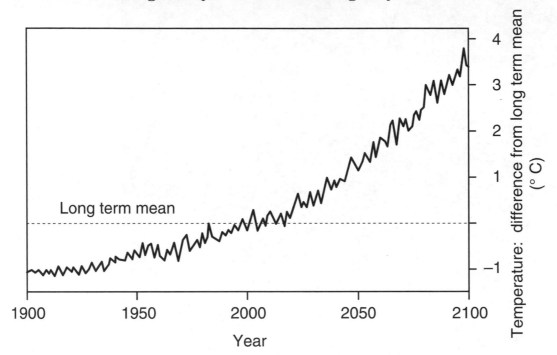

Marks

Question 2: Lithosphere

Study Diagram Q2.

Scree is a feature of both glaciated and limestone upland landscapes.

(*a*) **Describe** and **explain** the conditions **and** processes which encourage the formation of scree slopes.   **7**

Corries are landscape features in glaciated upland areas.

(*b*) With the aid of annotated diagrams, **explain** the processes involved in the formation of a corrie.   **9**

Diagram Q2: Scree

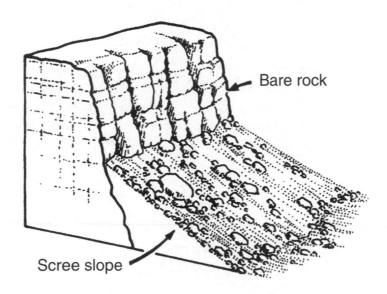

[Turn over

Marks

Question 3: Population Geography

Map Q3 shows the main origins of UK immigrants during 2005/2006.

(*a*)   **Describe** and **suggest reasons** for the patterns shown on Map Q3.   **10**

(*b*)   With reference to a migration flow you have studied, **describe** the impact on **either** the donor **or** receiving country.   **6**

Map Q3: Main Origins of UK Immigrants 2005/2006

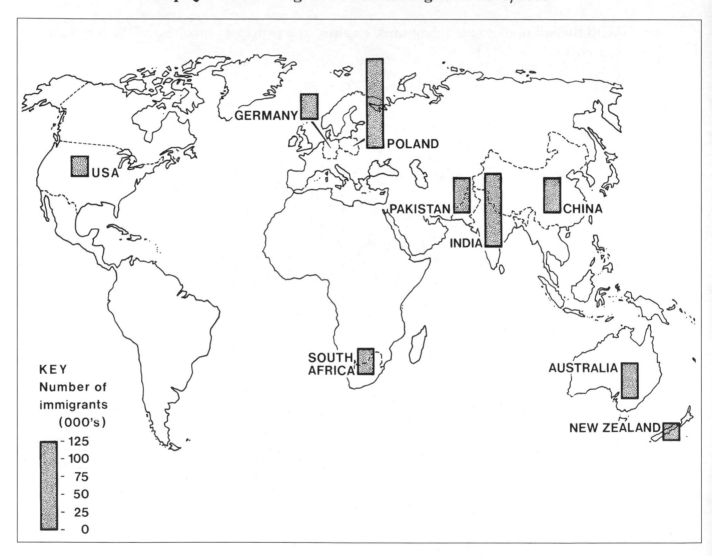

Marks

Question 4: Urban Geography

Study OS Map Extract number 1788/105: York (*separate item*), and Map Q4.

(*a*)   What **map evidence** suggests that the Central Business District of York lies within Area A?   **6**

(*b*)   For **either** Area B **or** Area C, **explain** the advantages of its location and environment for its residents.   **7**

(*c*)   Using map evidence, **explain** why the southward expansion of York into Area D may create land use conflicts.   **7**

Map Q4: Location of urban areas in York

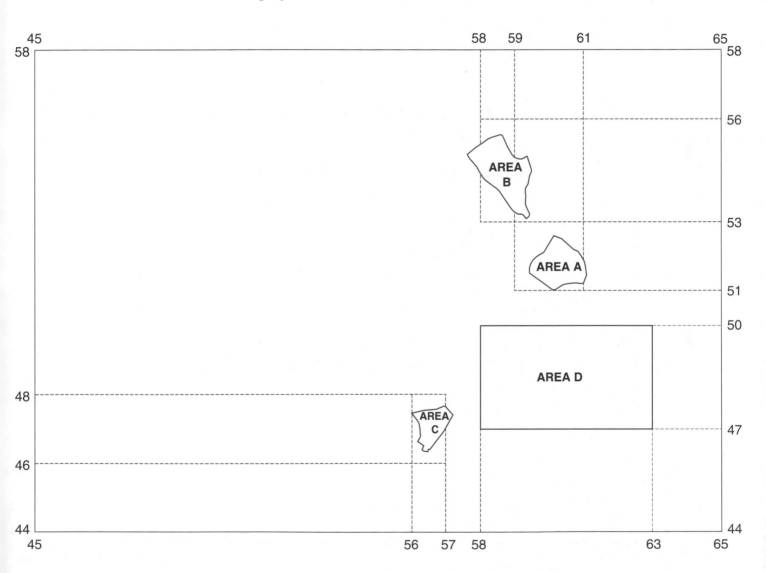

[Turn over

Marks

**SECTION B: Answer ONE question from this section,
ie either Question 5 or Question 6.**

Question 5: Hydrosphere

Study OS Map Extract number 1788/105: York (*separate item*).

(a) Meanders have formed on the River Nidd from GR 450542 to its confluence with the River Ouse GR 513578.

   Describe and **explain**, with the aid of a diagram or diagrams, how a meander is formed. **8**

(b) Study the OS Map Extract and Map Q5.

   "*The 2000 floods were the worst in York since records began and the River Ouse reached a height of 5·3 metres above its normal summer level.*"

   (BBC News, November 2000)

   With the aid of map evidence, **explain** the physical **and** human factors which may have contributed to the flooding in York after periods of extreme rainfall. **6**

Map Q5: Flooded areas of York, 2000

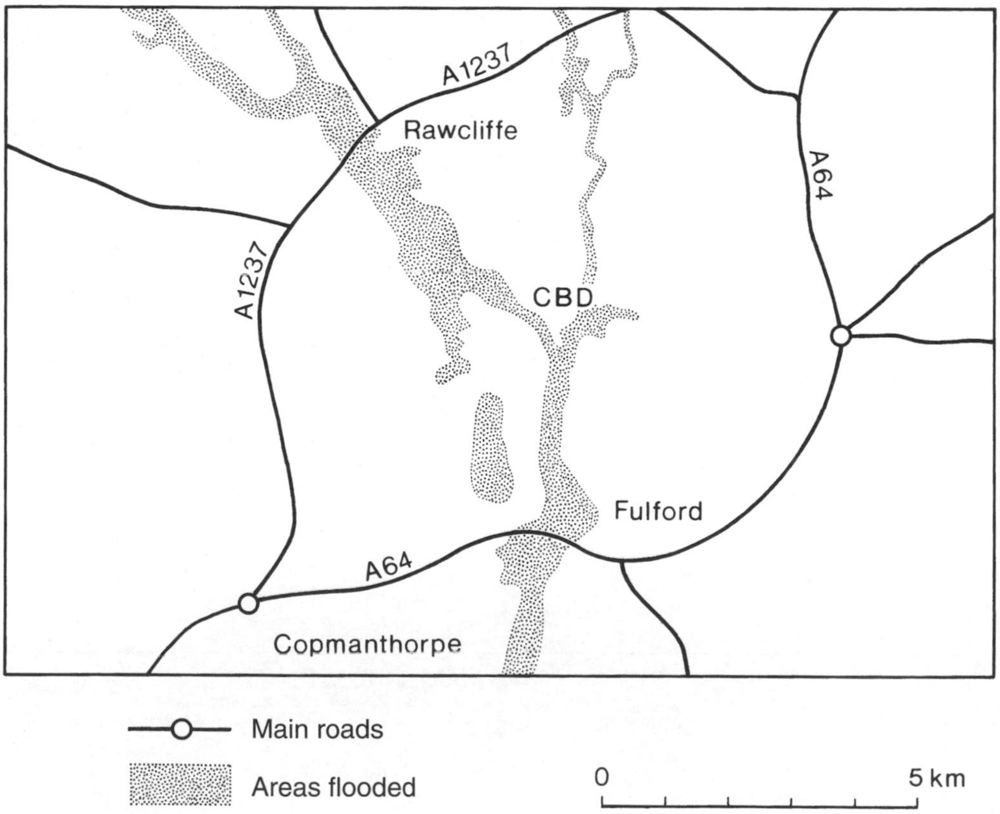

—○—	Main roads
(shaded)	Areas flooded

0 ————————— 5 km

Marks

DO NOT ANSWER THIS QUESTION IF YOU HAVE ALREADY ANSWERED QUESTION 5

Question 6: Biosphere

Study Diagram Q6 which shows some of the factors involved in vegetation succession on sand dunes.

Explain why there is a change in vegetation cover and species as you move inland from the beach. You should refer to named plant species in your answer.   **14**

Diagram Q6: Factors involved in sand dune succession

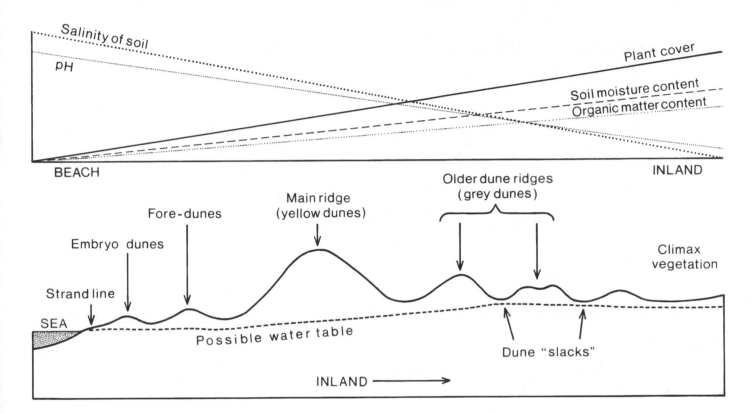

[Turn over

Marks

SECTION C: Answer ONE question from this section, ie either Question 7 or Question 8.

Question 7: Rural Geography

"Mechanisation has led to major changes in commercial arable farming."

(*a*)   (i)   **Suggest why** farmers have invested in increased mechanisation.

       (ii)   **Explain** the impact of increased mechanisation on the environment.   **8**

(*b*)   Study Diagram Q7 which shows some of the other recent changes in commercial arable farming.

       Describe and **explain two** of the changes shown in Diagram Q7.   **6**

Diagram Q7: Recent changes in commercial arable farming

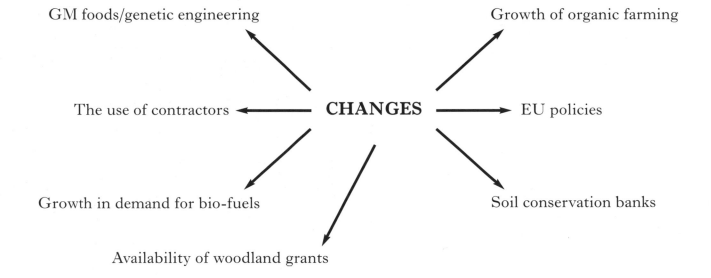

GM foods/genetic engineering   Growth of organic farming

The use of contractors   **CHANGES**   EU policies

Growth in demand for bio-fuels   Soil conservation banks

Availability of woodland grants

Marks

**DO NOT ANSWER THIS QUESTION IF YOU HAVE
ALREADY ANSWERED QUESTION 7**

Question 8: Industrial Geography

With reference to named examples within an area of industrial decline in the European Union you have studied:

 (i) **give reasons** for the industrial decline; and   **8**

(ii) **describe** the socio-economic impacts of the closure of such industries on the local population and the surrounding area.   **6**

[END OF QUESTION PAPER]

[BLANK PAGE]

X208/303

NATIONAL
QUALIFICATIONS
2010

MONDAY, 31 MAY
10.50 AM – 12.05 PM

GEOGRAPHY
HIGHER
Paper 2
Environmental
Interactions

Answer any **two** questions.

Write the numbers of the **two** questions you have attempted in the marks grid on the back cover of your answer booklet.

The value attached to each question is shown in the margin.

Credit will be given for appropriate maps and diagrams, and for reference to named examples.

Questions should be answered in sentences.

Note   The reference maps and diagrams in this paper have been printed in black only: no other colours have been used.

Marks

Question 1 (Rural Land Resources)

(*a*)   With the aid of annotated diagrams, **describe** and **explain** the physical features associated with the formation of coastal landscapes. You should refer to both erosion **and** deposition features in your answer.   **20**

(*b*)   For any named coastal area you have studied, **describe** how this landscape has provided a variety of socio-economic opportunities.   **10**

(*c*)   Study Diagram Q1A and Map Q1B.

One example of a land use conflict is the proposed leisure/housing development at the Menie Estate in Aberdeenshire. Part of this development takes place on a protected sand dune area designated as an SSSI. (SSSI = Site of Special Scientific Interest.)

Discuss the advantages **and** disadvantages of developments such as this on the local people and the environment.   **10**

(*d*)   For any named coastal **or** upland area you have studied, **describe** the measures taken to resolve environmental conflicts and **comment on** their effectiveness.   **10**

                                    (50)

Diagram Q1A: News Reports on the Proposals for the Menie Estate

"Business leaders have joined forces to urge the Scottish Government to give the go-ahead to US billionaire Donald Trump's plans for a golf resort . . .   Mr Trump hopes to build a resort featuring two championship golf courses, a five-star hotel, 950 holiday homes and 500 private houses at the Menie Estate in Aberdeenshire."

(*The Herald* 19/6/08)

"The value of Menie Links as part of the Foveran Links SSSI cannot be understated. It is the most dynamic, most rapidly moving and largest area of bare sand in this area of Scotland. It is quite simply the jewel in the crown of the SSSI areas of bare sand in this area of Scotland and therefore the jewel in the crown of the UK resource."

(Scottish Natural Heritage (SNH) expert—*The Herald* 19/6/08)

RSPB Scotland objected to the Trump International application because . . . the developer's own Environmental Statement acknowledges that there will be very significant adverse effects on habitats and biodiversity—the mobile dunes, which form one of the main qualifying features of the Foveran Links SSSI, will be destroyed.

(http://www.rspb.org.uk/ourwork/conservation/sites/scotland/menie.asp)

(RSPB = Royal Society for the Protection of Birds)

Question 1 – continued

Map Q1B: Proposed developments at the Menie Estate

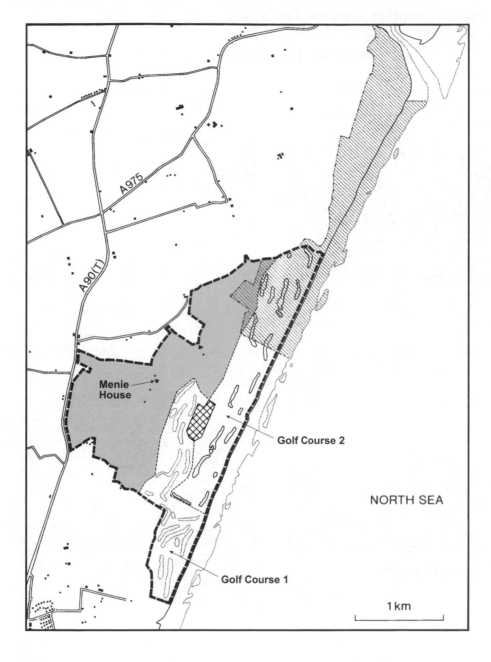

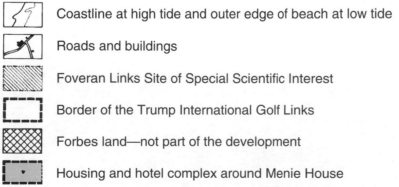

	Coastline at high tide and outer edge of beach at low tide
	Roads and buildings
	Foveran Links Site of Special Scientific Interest
	Border of the Trump International Golf Links
	Forbes land—not part of the development
	Housing and hotel complex around Menie House

Marks

Question 2 (Rural Land Degradation)

(*a*)   Study Diagram Q2A.

   Describe and **explain** the processes of soil erosion by water.   8

Diagram Q2A: Erosion by water

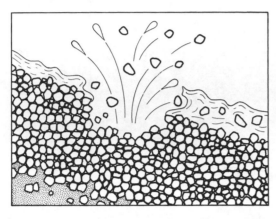

Rainsplash

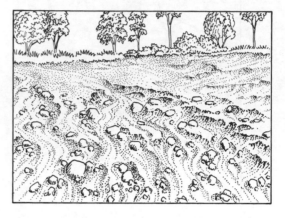

Sheet Erosion

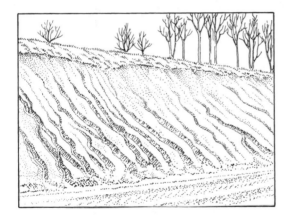

Rill Erosion

Gully Erosion

Marks

Question 2 – continued

(b) Study Table Q2B.

Describe and **explain** how human activities have caused land degradation in North America **and either** Africa north of the Equator **or** the Amazon Basin.   **16**

Table Q2B: Percentage of the agricultural land which has been degraded

Region	% degraded
North America	26
Africa	65
South America	45

(c) Referring to named locations in **either** Africa north of the Equator **or** the Amazon Basin, **describe** the impact of land degradation on the people and economy.   **10**

(d) Referring to named locations in North America you have studied:

(i) **describe** and **explain** the ways in which farmers have adjusted their farming methods to reduce the risk of soil erosion; and

(ii) **comment** on the effectiveness of these methods.   **16**

(50)

[Turn over

Marks

Question 3 (River Basin Management)

(*a*) *"The Myitsore hydro-electric project was started in 2008 to manage the flow of the Irrawaddy River in northern Myanmar (Burma)."*

Study Map Q3A and Diagrams Q3A, Q3B and Q3C.

(i) **Describe** and **account for** the pattern of river flow before the Myitsore Project started.

(ii) **Describe** and **explain** the need for water management in the Irrawaddy River in Myanmar. **16**

(*b*) For the Myitsore Dam **or** any named dam you have studied in Africa **or** North America **or** Asia, **describe** and **explain** the physical factors which should be considered when selecting the site for the dam and its associated reservoir. **10**

(*c*) **Describe** and **account for** the social, economic and environmental benefits **and** adverse consequences of a named water management project in Africa **or** North America **or** Asia. **24**

(50)

Diagram Q3A: Monthly discharge of the Irrawaddy River at Myitsore before the HEP scheme

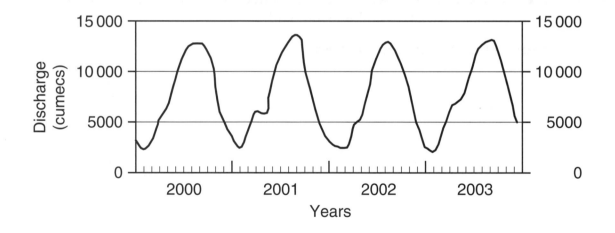

Question 3 – continued

Map Q3A: Irrawaddy River in Myanmar

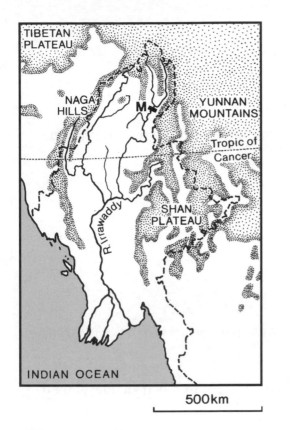

M↖ Myitsore Dam

▨ Land over 1000 metres

Diagram Q3B: Myitsore—Climate Graph

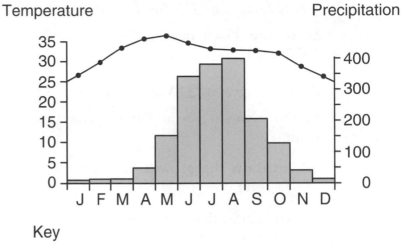

Key

●—● Temperature

▨ Precipitation

Diagram Q3C: Projected population change in Myanmar

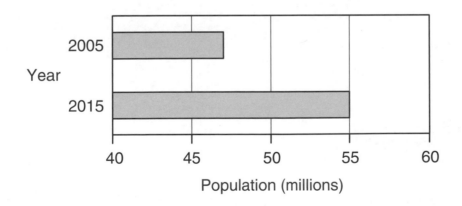

[Turn over

Marks

Question 4 (Urban Change and its Management)

(*a*) "*A megacity is defined as a city with over 10 million people.*"

Study Diagram Q4A (on *Page nine*).

Describe the changes in the number and world distribution of megacities from 1975 to 2015. **8**

(*b*) Study Diagrams Q4A and Q4B (on *Pages nine* and *ten*).

For Mexico City **or** any other named city which you have studied in a Developing Country:

(i) **explain** the growth of your chosen city in terms of rural push/urban pull factors; **10**

(ii) **describe** the socio-economic and environmental problems which have resulted from this rapid growth. **12**

(*c*) Study Map Q4 (on *Page ten*).

"*Urban sprawl has been seen as a problem since the 1930s and regions such as South-East England have come under increasing pressure.*"

Referring to London **or** any other named city you have studied in a Developed Country:

(i) **explain** the reasons for urban sprawl;

(ii) **outline** the problems caused by this growth;

(iii) select **one** problem identified in part (ii) above and **explain** the ways in which the city has tried to resolve this problem. **20**

(50)

Question 4 – continued

Diagram Q4A: The Growth of Megacities 1975–2015

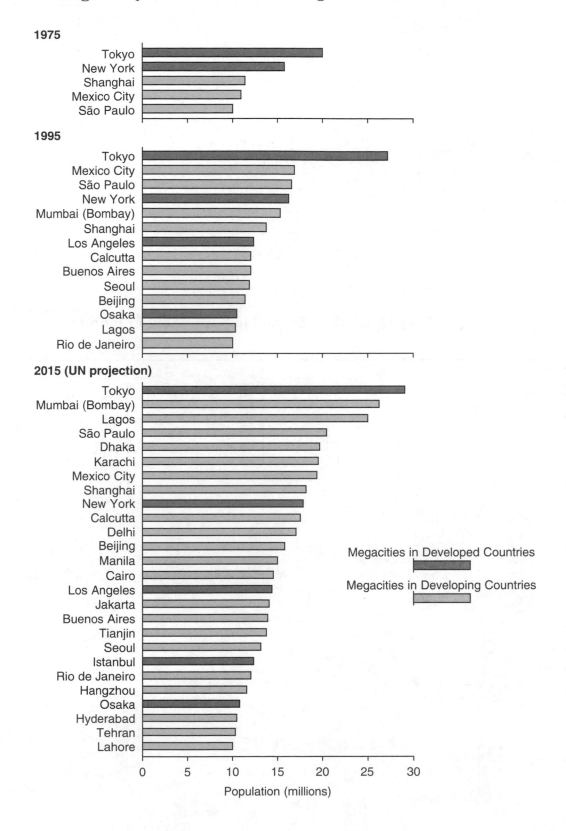

Question 4 – continued

Diagram Q4B: Changes in balance of urban/rural population in Mexico

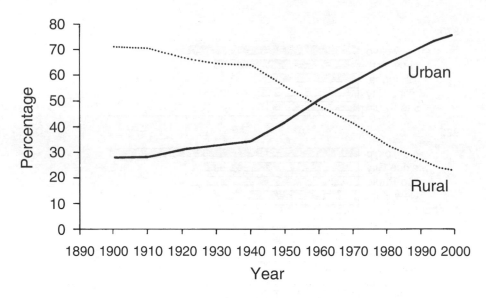

Map Q4: Urban Sprawl in South-East England

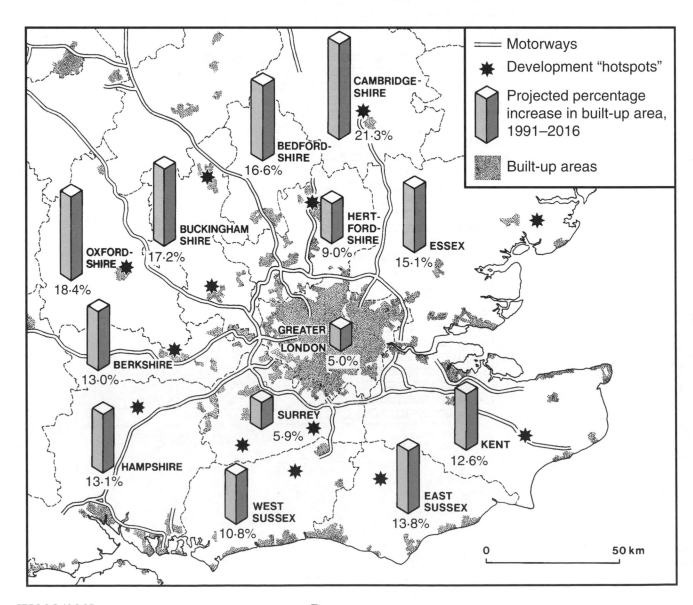

[Turn over for Question 5 on *Page twelve*

Marks

Question 5 (European Regional Inequalities)

(*a*) Study Map Q5A.

"Convergence Regions are areas designated as requiring most financial assistance across the European Union (EU)."

(i) **Describe** the distribution of the Convergence Regions.   8

(ii) In 2008, the EU budget for promoting growth across the least developed regions was 47 billion Euros. **Discuss** ways in which less prosperous regions can receive help from the EU.   10

(*b*) *"The **North-South divide** refers to the economic and cultural differences between southern England and the rest of the United Kingdom."*

Study Map Q5B and Table Q5.

(i) To what extent does the data provide evidence of regional inequalities within the UK?   10

(ii) **Describe** and **explain** the physical and human factors that have led to the regional inequalities within the UK.   14

(iii) **Describe** the steps taken by the UK government agencies to reduce regional inequalities.   8

(50)

Map Q5A: EU Convergence Regions

▨ Convergence Regions receiving most financial aid

Question 5 – continued

Map Q5B: UK statistical regions

Table Q5: UK average values

	Population change 1996–2006 %	Average house prices (Nov 2008) £1000	Gross disposable household income (2006) (UK average =100)	Working age population with no qualifications (%) (2006)
UK average	4·3	203	100	13
Scotland	0·0	160	95	13
Northern Ireland	5·1	226	87	22
Wales	2·4	158	89	17
NW England	0·0	158	92	15
NE England	−1·1	148	86	14
West Midlands	2·3	176	91	17
Yorks & Humber	2·7	157	93	15
East Midlands	5·5	166	95	13
East England	6·6	204	107	12
SE England	6·7	271	113	9
London	7·4	382	120	12
SW England	6·9	229	100	9

Marks

Question 6 (Development and Health)

(*a*)   **Suggest reasons** for the wide variations in development which exist **between** Developing Countries. You should refer to named countries you have studied.   **12**

(*b*)   Study Table Q6A, and Maps Q6A, Q6B, Q6C and Q6D.

"Life expectancy in Chad is only 47 years."

Suggest the physical **and** human factors which may have led to this low life expectancy.   **12**

(*c*)   Study Map Q6C.

Chad and many other developing countries have been affected by water-related diseases including malaria, cholera and bilharzia/schistosomiasis.

Select **one** of the above diseases.

(i)   **Describe** the physical **and** human factors which put people at risk of contracting the disease.   **8**

(ii)   **Describe** and **explain** the measures that can be taken to combat the disease.   **14**

(iii)   **Explain** the benefits to a Developing Country of controlling the disease.   **4**

(50)

Table Q6A: Selected development indicators for Chad

Indicator	
GDP per capita ($US)	1500
Birth rate per 1000	42
Infant mortality per 1000 live births	100
% land surface for arable farming	3
Adult literacy rate (%)	25
% population with HIV/AIDS	4·8

Question 6 – continued

Map Q6A: Map of Chad

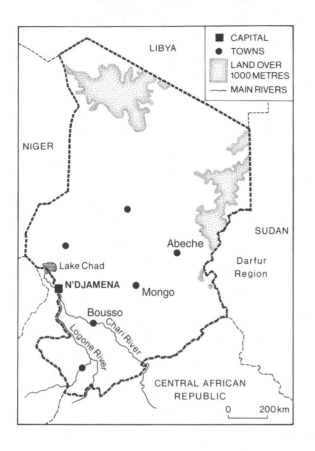

Map Q6B: Bio-climatic zones of Chad

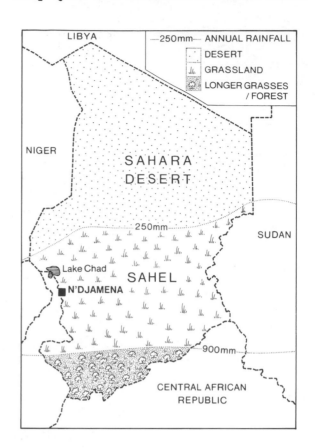

Map Q6C: Disease in Chad

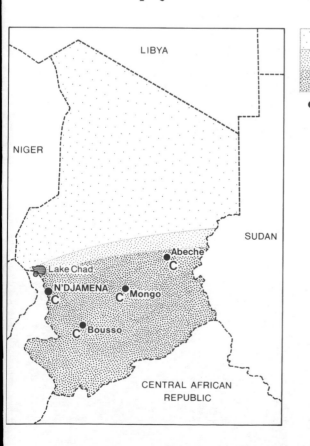

Map Q6D: Location of Chad within Africa

[END OF QUESTION PAPER]

[BLANK PAGE]

HIGHER

2011

[BLANK PAGE]

X208/301

NATIONAL QUALIFICATIONS 2011	TUESDAY, 24 MAY 9.00 AM – 10.30 AM	GEOGRAPHY HIGHER Paper 1 Physical and Human Environments

Six questions should be attempted, namely:

all four questions in **Section A** (Questions 1, 2, 3 and 4);

one question from **Section B** (Question 5 **or** Question 6);

one question from **Section C** (Question 7 **or** Question 8).

Write the numbers of the **six** questions you have attempted in the marks grid on the back cover of your answer booklet.

The value attached to each question is shown in the margin.

Credit will be given for appropriate maps and diagrams, and for reference to named examples.

Questions should be answered in sentences.

Note   The reference maps and diagrams in this paper have been printed in black only: no other colours have been used.

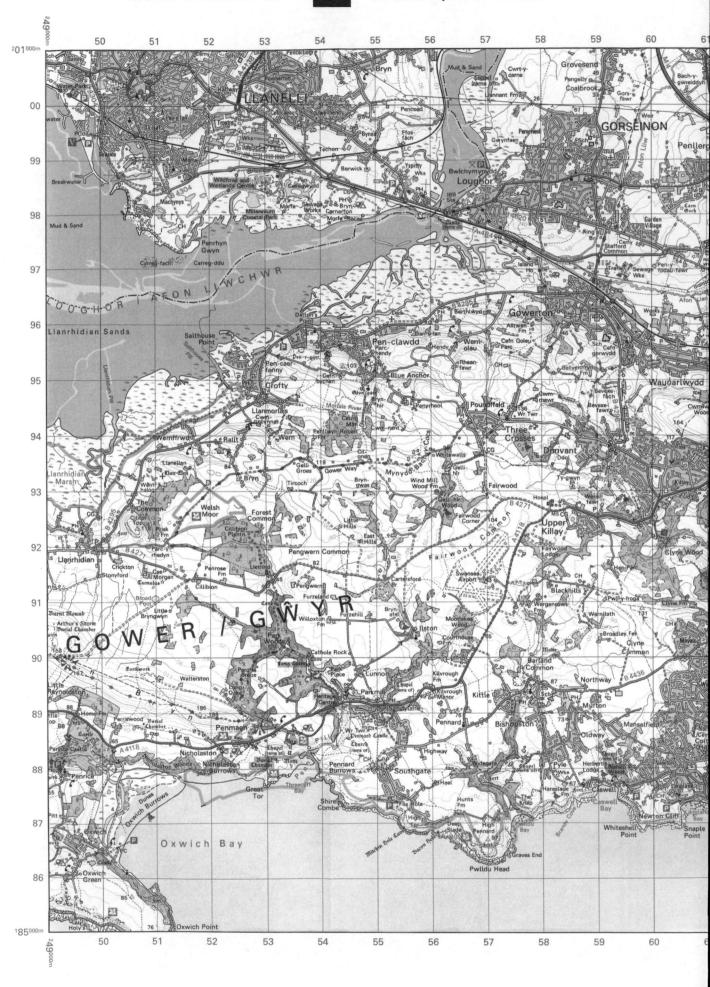

Scale 1: 50 000
2 centimetres to 1 kilometre (one grid square)

Marks

Question 1: Atmosphere

Study Maps Q1A.

(*a*) **Describe** the origin, nature and characteristics of the Maritime Tropical and
Continental Tropical air masses.   **6**

Maps Q1A: Location of selected air masses and the ITCZ in January and July

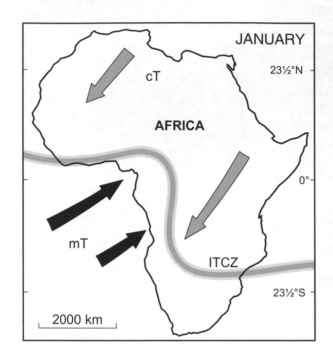
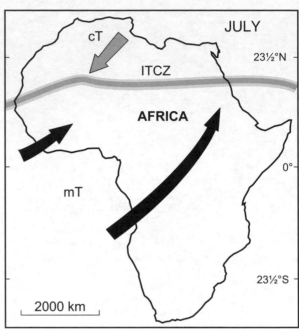

Key:

mT   =   Maritime Tropical   cT   =   Continental Tropical

ITCZ = Inter Tropical Convergence Zone

Question 1: Atmosphere (continued)

Study Maps Q1A and Q1B and Diagram Q1.   *Marks*

(b)   **Describe** and **explain** the variation in rainfall within West Africa.   **12**

Map Q1B: West Africa

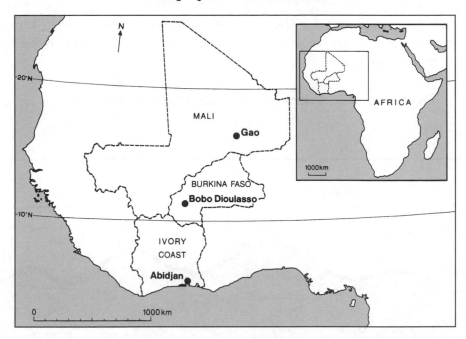

Diagram Q1: Average Monthly Rainfall/Days with Precipitation

Gao: total precipitation—200 mm

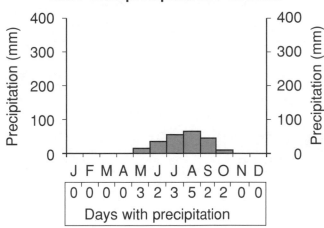

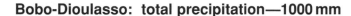

Bobo-Dioulasso: total precipitation—1000 mm

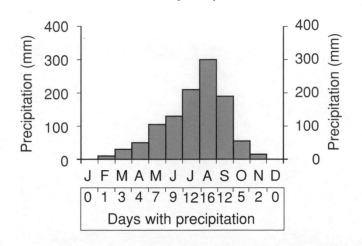

Abidjan: total precipitation—1700 mm

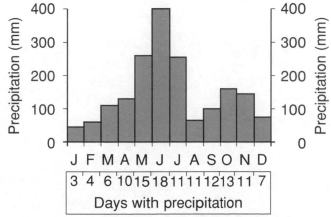

Marks

Question 2: Biosphere

(*a*)   **Explain** fully what is meant by the term climax vegetation.   **5**

Study Diagram Q2.

(*b*)   **Describe** and **give reasons** for the changes in plant types likely to be observed across the transect as you move inland from the coast. You should refer to named plant species in your answer.   **13**

Diagram Q2: Transect of Sand Dune System

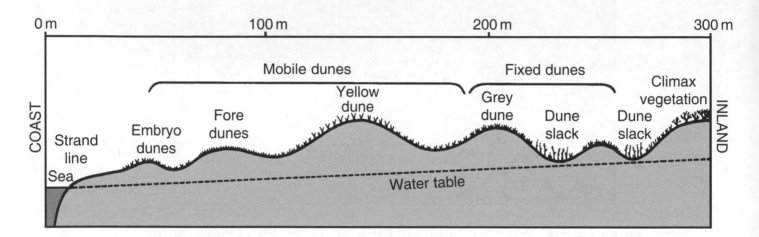

Marks

Question 3: Rural Geography

Study Map Q3.

Referring to a named area in **either** a shifting cultivation **or** an intensive peasant agricultural system:

(i) **describe** and **explain** the main features of your chosen farming landscape;   8

(ii) **describe** the recent changes that have taken place; **and**

   discuss the impact of these changes on the people and their environment.   10

Map Q3: Generalised distribution of selected agricultural systems

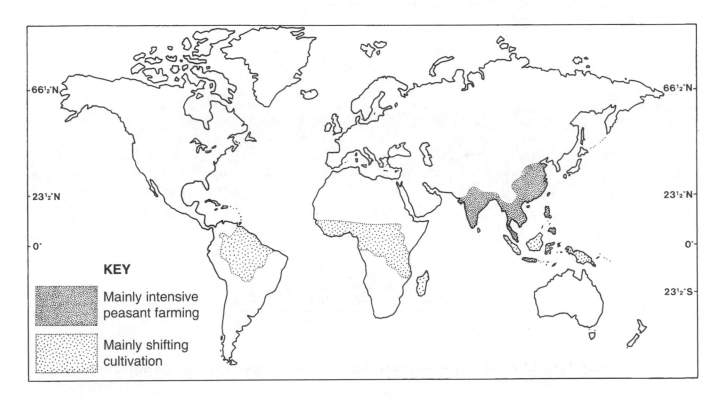

[**Turn over**

Question 4: Industry

Study OS Map Extract number 1882/159: Swansea (*separate item*), and Maps Q4A and Q4B.

(*a*) Using map evidence, **describe** and **explain** the physical **and** human factors that have encouraged industry to locate in areas A and B.   **10**

(*b*) Study Map Q4B.

Explain ways in which the European Union **and** national government can create industrial regeneration in South Wales, **or** any other named industrial concentration in the EU which you have studied.   **8**

Map Q4A: Location of industrial areas in Swansea

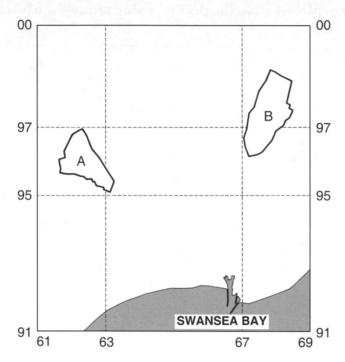

Map Q4B: South Wales—Location of new industrial developments

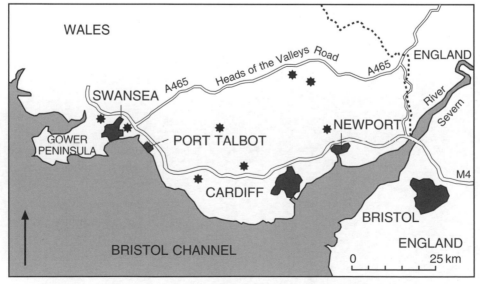

* Location of industrial developments since 1950

Marks

**SECTION B: Answer ONE question from this section,
ie either Question 5 or Question 6.**

Question 5: Lithosphere

Study OS Map Extract number 1882/159: Swansea (*separate item*).

(*a*)   Using map evidence, **identify** the features of coastal erosion from GR 513851 (Oxwich Point) to GR 636871 (Mumbles Head).   **6**

(*b*)   With the aid of annotated diagrams, **explain** the formation of **one** of the erosional features described in part (*a*).   **8**

[Turn over

Marks

**DO NOT ANSWER THIS QUESTION IF YOU HAVE
ALREADY ANSWERED QUESTION 5**

Question 6: Hydrosphere

(*a*)   Study Diagram Q6A.

*"A drainage basin is an open system with four elements—**inputs**, **storage**, **transfers** and **outputs**."*

Describe the movement of water within a drainage basin with reference to the four elements above.

7

Diagram Q6A: A Drainage Basin

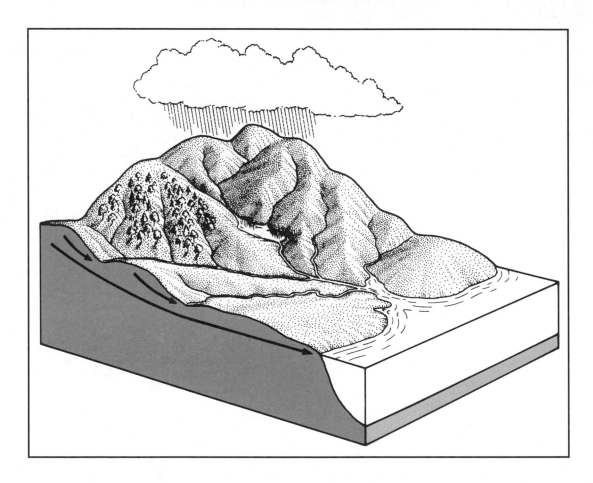

Marks

Question 6: Hydrosphere (continued)

(*b*) **Describe** and **explain** the changing river levels on the River Thaw at Cowbridge on 26 July 2007.

7

Diagram Q6B: Flood Hydrograph for the River Thaw at Cowbridge, 26 July 2007

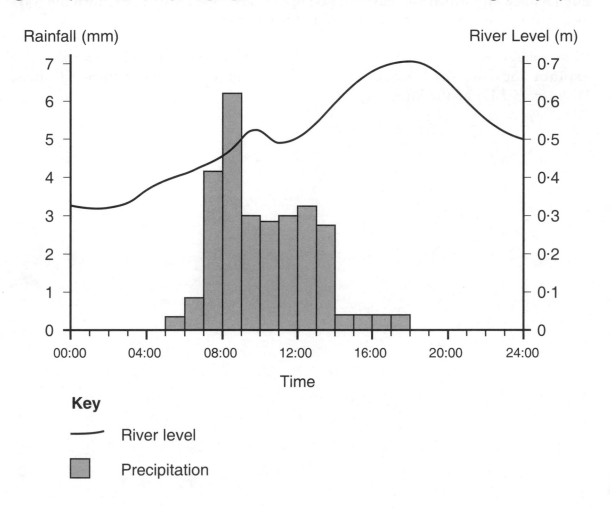

[X208/301]

[Turn over

Marks

**SECTION C: Answer ONE question from this section,
ie either Question 7 or Question 8.**

Question 7: Urban Geography

(*a*) For a named city which you have studied in the Developed World, **explain** the ways
in which its site **and** situation contributed to its growth.   **6**

(*b*) With reference to any Developed World City you have studied, **describe** and
explain the land use changes in recent years in **either** the Central Business
Disctrict (CBD) **or** the inner city.   **8**

Marks

**DO NOT ANSWER THIS QUESTION IF YOU HAVE
ALREADY ANSWERED QUESTION 7**

Question 8: Population Geography

"Apart from 1941, during the Second World War, the UK has carried out a census every 10 years since 1801."

(*a*) **Describe** how the UK gathers population data **between** these censuses and **explain** why it is important for countries to obtain accurate population data.   5

(*b*) Giving named examples, **explain** why carrying out a census may be more difficult and the results less reliable in Developing Countries than in Developed Countries such as the UK.   9

[END OF QUESTION PAPER]

[BLANK PAGE]

X208/303

NATIONAL
QUALIFICATIONS
2011

TUESDAY, 24 MAY
10.50 AM – 12.05 PM

GEOGRAPHY
HIGHER
Paper 2
Environmental
Interactions

Answer any **two** questions.

Write the numbers of the **two** questions you have attempted in the marks grid on the back cover of your answer booklet.

The value attached to each question is shown in the margin.

Credit will be given for appropriate maps and diagrams, and for reference to named examples.

Questions should be answered in sentences.

Note The reference maps and diagrams in this paper have been printed in black only: no other colours have been used.

Marks

Question 1 (Rural Land Resources)

(a)   Study Map Q1A.

Describe and **explain**, with the aid of annotated diagrams, the formation of the main features of glacial erosion in the Lake District **or** any other glaciated upland area which you have studied.   **18**

(b)   Study Diagram Q1.

With reference to the area around Coniston Valley, **or** any other upland area you have studied, **explain** the social and economic opportunities created by the landscape.   **10**

(c)   Study Map Q1B.

The Lake District and Snowdonia are both areas of outstanding glaciated scenery. **Explain** why these two National Parks attract widely differing numbers of visitors.   **6**

(d)   For the Lake District **or** any other upland **or** coastal area you have studied:

(i)   **explain** the environmental conflicts that may occur due to an influx of visitors. (You should refer to specific named examples within your chosen area);   **10**

(ii)   for **one** of the conflicts explained in part (i), **describe** the solutions to this conflict and comment on their effectiveness.   **6**

(50)

Map Q1A: The Lake District

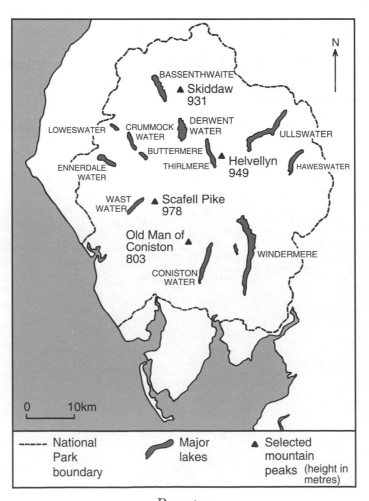

Question 1 – continued

Diagram Q1: Sketch Cross-section of the Coniston Valley in the Lake District

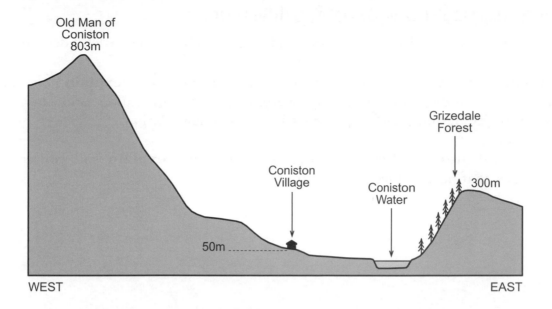

Map Q1B: Major Roads and Settlements around the Lake District and Snowdonia

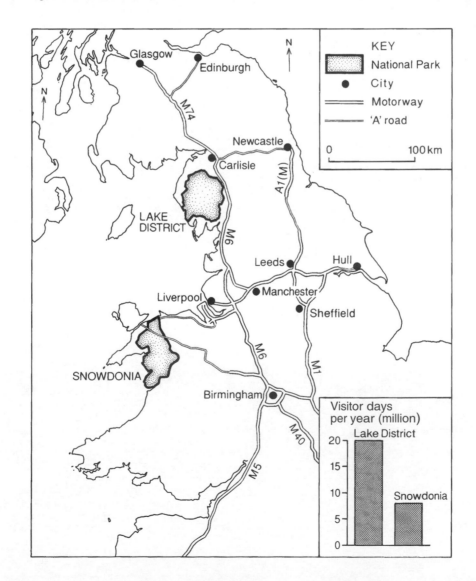

Marks

Question 2 (Rural Land Degradation)

(*a*) **Describe** the processes of water and wind erosion which lead to soil degradation.   **12**

(*b*) Study Map Q2 and Diagrams Q2A, Q2B and Q2C.

   Describe and **explain** why the climate of Niger has led to severe land degradation.   **8**

(*c*) For **either** Africa north of the equator, **or** the Amazon Basin, **explain** how human activities including deforestation, overgrazing, overcultivation and any other inappropriate farming techniques have led to land degradation.   **16**

(*d*) For named areas of North America, **describe** and **explain** soil conservation strategies that have reduced land degradation.   **14**

(50)

Question 2 – continued

Map Q2: Location of Niger

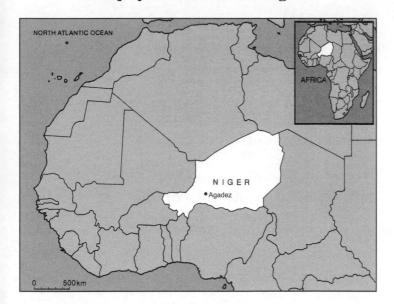

Diagram Q2A: Climate of Agadez, Niger

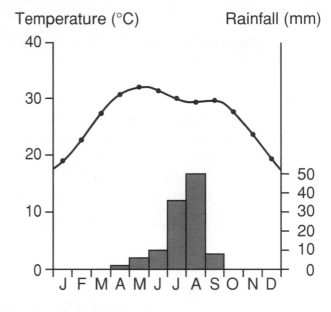

Key

⌒ Temperature

▮ Precipitation

Diagram Q2B: Average daily sunshine hours, Niger

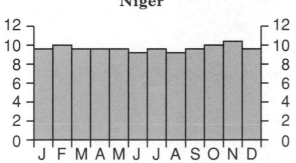

Diagram Q2C: Rainfall Variability in Niger

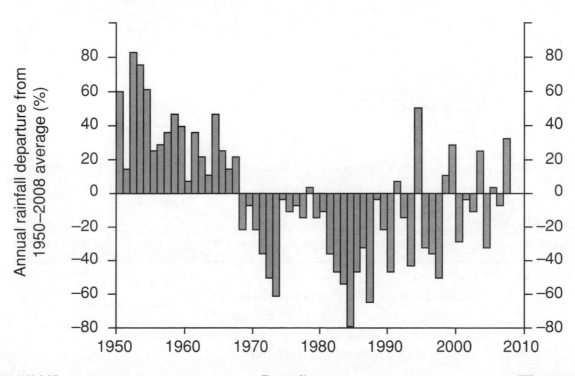

Marks

Question 3 (River Basin Management)

(*a*) Study Map Q3, Graphs Q3A and Q3B, and Table Q3.

Explain why there is a need for water management in the Malaysian owned area of the island of Borneo.   **12**

(*b*) For the Bakun Dam **or** any water control project you have studied in Africa **or** North America **or** Asia, **explain** the physical factors that should be considered when selecting sites for the dam(s) and associated reservoir(s).   **10**

(*c*) (i) **Describe** and **explain** the social, economic and environmental **benefits** of a named major water control project in Africa **or** North America **or** Asia.   **16**

(ii) Comment on any **problems** caused by your chosen water control project.   **12**

(50)

Map Q3: Proposed Bakun Dam, Malaysia

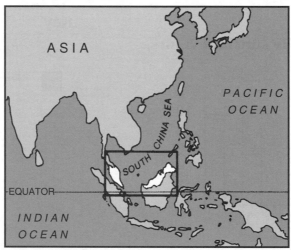

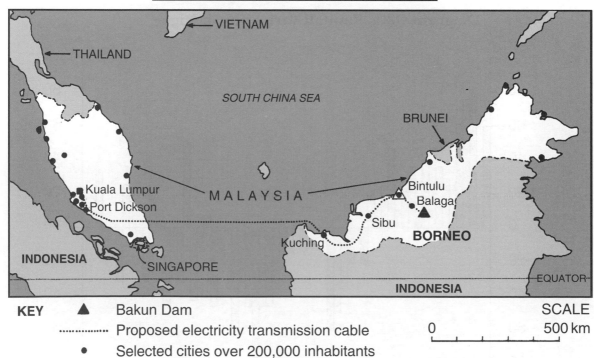

KEY   ▲   Bakun Dam

....   Proposed electricity transmission cable

•   Selected cities over 200,000 inhabitants

■   Kuala Lumpur (capital city), over 1 million inhabitants

▲   Proposed aluminium smelter (Bintulu)

SCALE

0 ———————— 500 km

Question 3 – continued

Graph Q3A: Balaga Climate Graph   **Graph Q3B: Population of Malaysia (millions)**

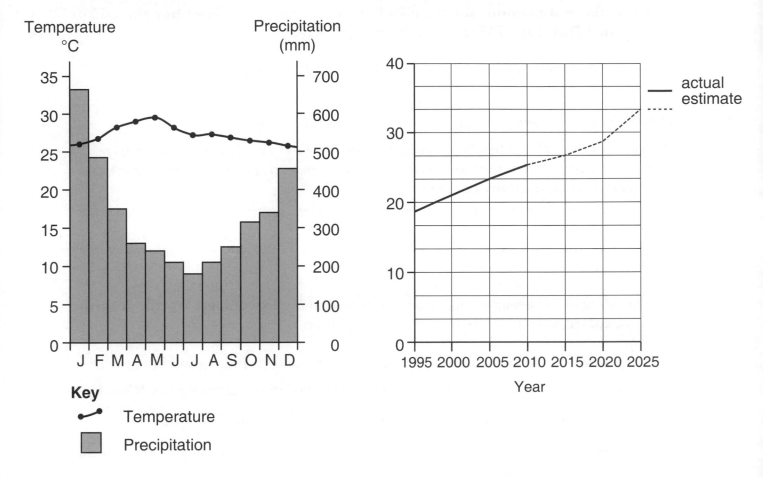

Table Q3: Indicators of Development—Malaysia

Urban population (2008)	70%
Rate of urbanisation (annual rate of change from 2005–10 estimate)	3%
Electricity production (2007)	102·9 billion kwh (kilowatt hours)
Electricity consumption (2010 estimate)	99·8 billion kwh
Electricity exports (2010 estimate)	2·3 billion kwh

[Turn over

Marks

Question 4 (Urban Change and its Management)

(*a*)   Study Map Q4A (on *Page nine*).

     Describe and **account for** the distribution of major cities in **either** the UK **or** any other **Developed World** country that you have studied.   **10**

(*b*)   Study Map Q4B (on *Page ten*) and Table Q4 (on *Page eleven*).

> "*A Games Like No Other—Glasgow 2014 Commonwealth Games.*
>
> *Glasgow will host the 20th Commonwealth Games where 71 countries will compete in 17 sports at various venues across the city. New facilities will involve massive investment and regeneration, particularly in the east end of the city.*"

     Discuss the advantages **and** disadvantages of this development for the residents of the East End of Glasgow.   **10**

(*c*)   For Glasgow, **or** a named city you have studied in the **Developed World**, **describe** and **explain** why it suffers from traffic congestion.   **10**

(*d*)   Many cities in the **Developing World** are experiencing rapid population growth.

     With reference to a named city that you have studied in the **Developing World**:

     (i)   **describe** and **explain** the problems caused by this rapid growth;   **12**

     (ii)   **describe** the methods the residents and local authorities might use to tackle these problems.   **8**

     (50)

Question 4 – continued

Map Q4A: Largest Cities in the UK

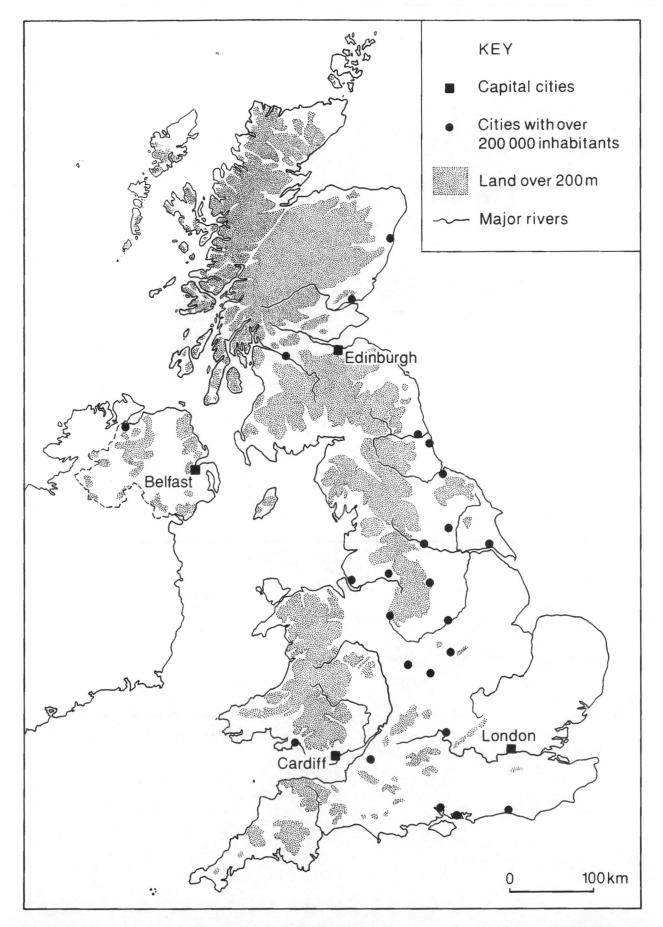

Map Q4A: Largest Cities in the UK

Question 4 – continued

Map Q4B: Selected Commonwealth Games Venues and Events Maps

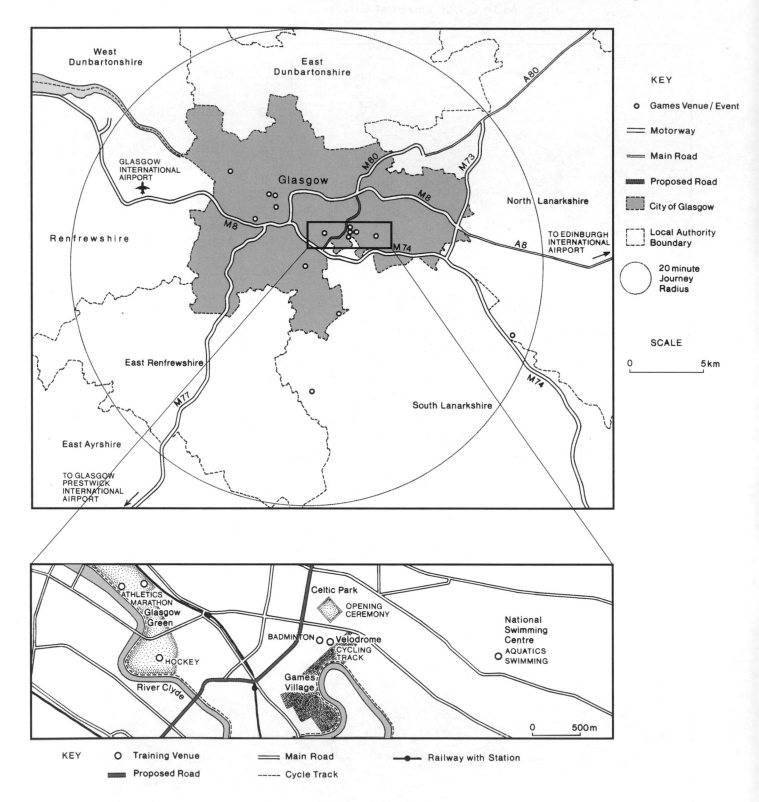

Question 4 – continued

Table Q4: Games Statistics

Public investment	£298 million (80% from the Scottish Government and 20% from Glasgow City Council)
Jobs	1000 for Glasgow, 1200 for Scotland
Net economic gain	£81 million
Volunteers	15,000 to be trained

[Turn over

Marks

Question 5 (European Regional Inequalities)

(a) Study Table Q5A.

> *"The European Union in 2009 had 27 member states and a waiting list of countries who had applied to join; including Turkey, Croatia, the Former Yugoslav Republic of Macedonia and Iceland."*

Suggest reasons why countries may wish to become members of the European Union.

10

(b) Study Map Q5 and Table Q5B.

Italy is often described as having a "North-South Divide". With reference to specific provinces and data provided in the table, **comment** on the accuracy of this statement.

12

(c) For Italy **or** any other country of the European Union which has marked differences in economic development between regions, **describe** the physical **and** human factors that have contributed to such regional differences.

16

(d) For the country chosen in part (c):

  (i) **discuss** ways in which less prosperous regions can receive help from their national government to overcome such problems; and

  (ii) **comment** on the effectiveness of these strategies.

12

(50)

Table Q5A: GNP per Capita data for selected EU countries

EU Country	Year of joining	GNP per capita (PPP*) 2004	GNP per capita (PPP*) 2008	% Increase in GNP per capita (PPP*)
UK	1973	$31,430	$36,130	15
France	1957	$29,460	$34,400	17
Slovakia	2004	$14,480	$21,300	47
Poland	2004	$12,730	$17,310	36

PPP* = Purchasing Power Parity

Question 5 – continued

Map Q5: Italy—GNP Per Capita by Provinces

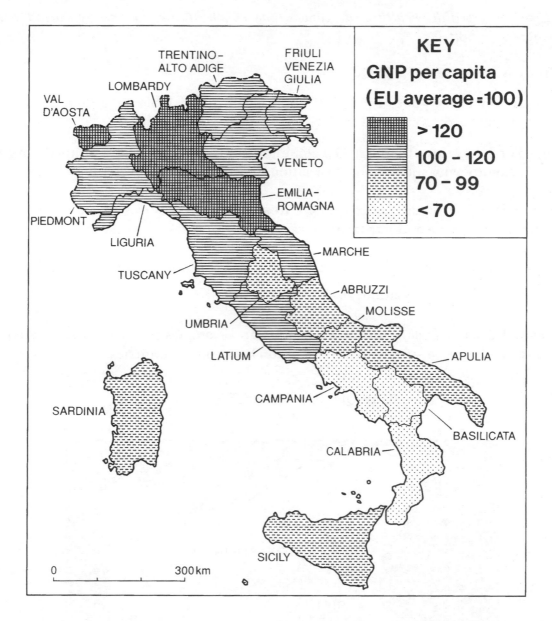

Table Q5B: Italy—Percentage of total employment in selected sectors in North and South

Employment Sector	Percentage in north	Percentage in south
Industry	78	22
Commerce	65	35
Services	59	41

[Turn over

Marks

Question 6 (Development and Health)

(a) *"Levels of wealth, health and economic development are not evenly spread within individual countries."*

Study Map Q6 and Table Q6.

   (i) In what ways does the information given in the table suggest that the eight provinces of Kenya are at different levels of development?   **10**

   (ii) For Kenya **or** any other Developing Country that you have studied, **suggest reasons** why such regional variations exist **within** a Developing Country.   **10**

(b) For malaria, **or** bilharzia, **or** cholera, **describe** the measures that can be taken to combat the disease and **explain** the varying effectiveness of these measures.   **18**

(c) *"The ultimate goal of Primary Health Care is better health for all."*

(World Health Organisation)

Describe some specific Primary Health Care strategies that you have studied and **explain** why these strategies are suited to Developing Countries.   **12**

(50)

Map Q6: Provinces of Kenya

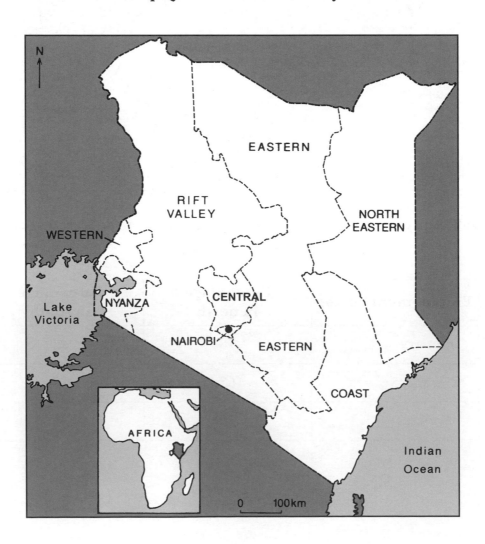

Question 6 – continued

Table Q6: Selected socio-economic indicators of development for Kenya's provinces

Province of Kenya	% Females with no education	% Males with no education	% Population below the poverty line	% Children aged 12–23 months without all vaccinations
North Eastern	87	66	64	92
Coast	38	23	58	36
Eastern	20	14	58	37
Central	12	8	31	24
Nairobi	10	8	44	40
Rift Valley	28	22	48	46
Western	18	11	61	52
Nyanza	18	10	65	64

[END OF QUESTION PAPER]

[BLANK PAGE]

HIGHER

2012

[BLANK PAGE]

X208/12/01

NATIONAL QUALIFICATIONS 2012	TUESDAY, 8 MAY 9.00 AM – 10.30 AM	GEOGRAPHY HIGHER Paper 1 Physical and Human Environments

Six questions should be attempted, namely:

all four questions in **Section A** (Questions 1, 2, 3 and 4);

one question from **Section B** (Question 5 **or** Question 6);

one question from **Section C** (Question 7 **or** Question 8).

Write the numbers of the **six** questions you have attempted in the marks grid on the back cover of your answer booklet.

The value attached to each question is shown in the margin.

Credit will be given for appropriate maps and diagrams, and for reference to named examples.

Questions should be answered in sentences.

Note The reference maps and diagrams in this paper have been printed in black only: no other colours have been used.

Extract No 1940/115

1:50 000 Scale
Landranger Series

Four colours should appear above; if not then please return to the invigilator.
Four colours should appear above; if not then please return to the invigilator.

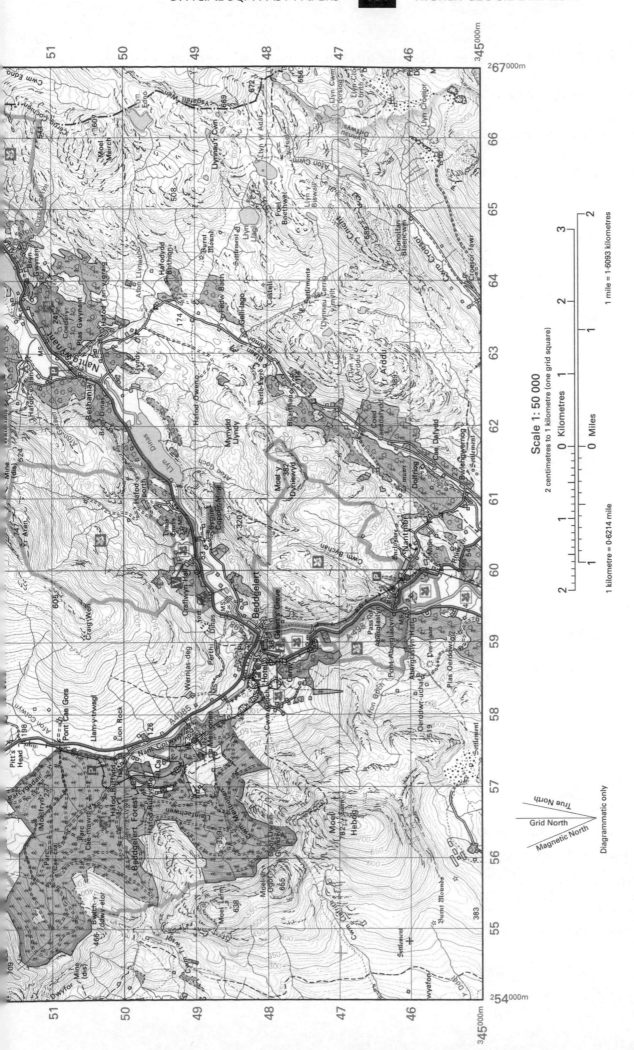

Scale 1 : 50 000

2 centimetres to 1 kilometre (one grid square)

1 kilometre = 0·6214 mile

1 mile = 1·6093 kilometres

Diagrammatic only

True North
Grid North
Magnetic North

Extract produced by Ordnance Survey 2011. Licence: 100035658
© Crown copyright 2010. All rights reserved.

Marks

SECTION A: Answer ALL four questions from this section.

Question 1: Lithosphere

Study OS Map Extract number 1940/115: Snowdon (*separate item*), **and** Map Q1.

(*a*)   **Describe** the evidence which shows that Area A, shown on Map Q1, has been affected by the processes of glacial erosion.

You should refer to specific named features and make use of grid references.   **12**

Map Q1: Snowdon

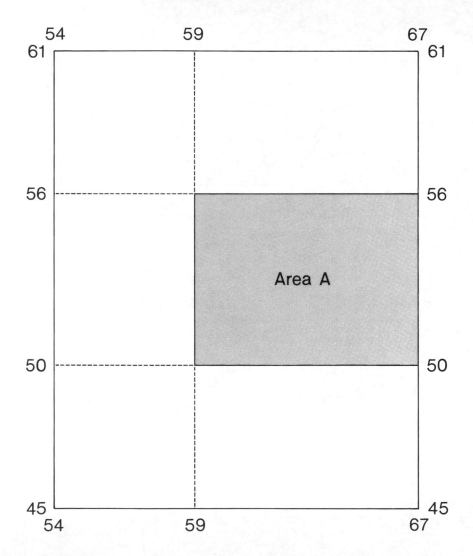

(*b*)   **Explain**, with the aid of an annotated diagram or diagrams, how **one** of the following features of glacial deposition is formed:

*   terminal moraine

*   esker

*   drumlin.   **6**

Marks

Question 2: Hydrosphere

(*a*)   Study Diagram Q2A.

Describe and **explain** how human activities, such as those shown on Diagram Q2A, can impact on the hydrological cycle.   **10**

Diagram Q2A: Human activities affecting the hydrological cycle

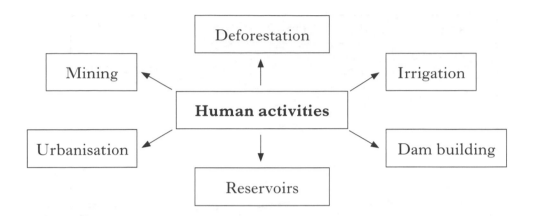

(*b*)   Study Diagram Q2B.

Explain, with the aid of an annotated diagram or diagrams, how **one** of the lower course river features, shown on Diagram Q2B, is formed.   **8**

Diagram Q2B – Lower course river features

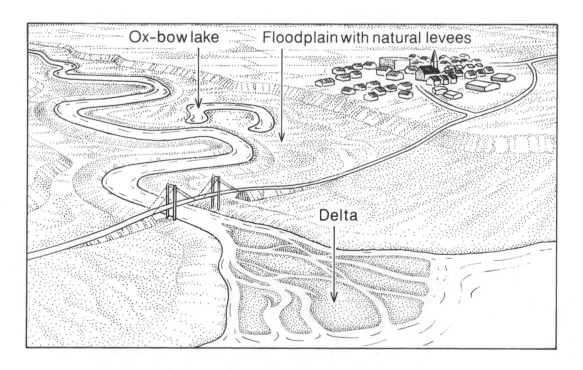

[Turn over

Marks

Question 3: Population

(*a*)   Study Diagram Q3A.

Describe and **suggest reasons** for the changes that take place in **Stages 1 to 3** of the Demographic Transition Model.

10

Diagram Q3A: The Demographic Transition Model

Stage 1	Stage 2	Stage 3	Stage 4	Stage 5

Total population

Death rate

Birth rate

(*b*)   Study Diagram Q3B.

Describe the changes which have taken place in Scotland's population structure and **suggest problems** that the government may face as a result of these changes.

8

Diagram Q3B: Scotland's population 2012 (Change from 1974 in brackets)

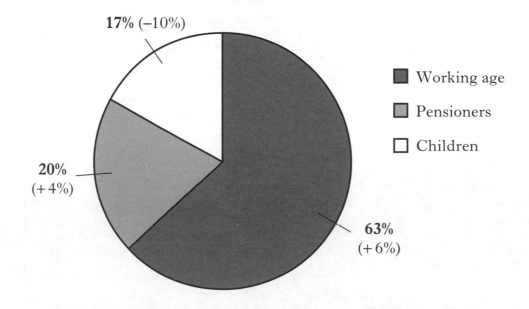

17% (−10%)

20% (+4%)

63% (+6%)

■ Working age

■ Pensioners

☐ Children

Marks

Question 4: Industrial Geography

Study Diagram Q4 **and** Map Q4.

With reference to **named examples** within an area of industrial decline in the European Union which you have studied:

(*a*)   **explain** reasons for industrial decline in areas like Teesside;   **8**

(*b*)   **describe** and **explain** the impact of industrial closures on people, the local economy and the environment in the surrounding area.   **10**

Diagram Q4: Newspaper Extract

> *"Workers in one of the UK's biggest steel-making areas were dealt a savage pre-Christmas blow today with news that a giant plant in Teesside is to be closed with the loss of 1700 jobs. Job losses leave Corus workers devastated."*
>
> Press Association December 2009

Map Q4: Corus Steelworks closures in England and Wales in 2009

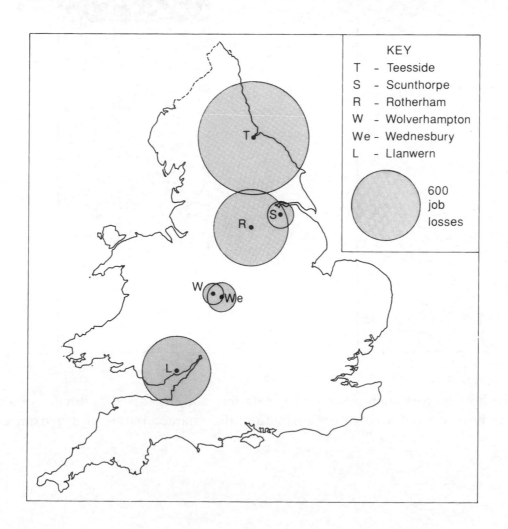

KEY

T   - Teesside
S   - Scunthorpe
R   - Rotherham
W   - Wolverhampton
We - Wednesbury
L   - Llanwern

600 job losses

Marks

SECTION B: Answer ONE question from this section, ie either Question 5 or Question 6.

Question 5: Biosphere

(a)   Study Diagram Q5.

Choose **one** of the soil profiles.

Describe the characteristics of the soil, including horizons, colour, soil biota, texture and drainage.   **6**

Diagram Q5: Selected soil profiles

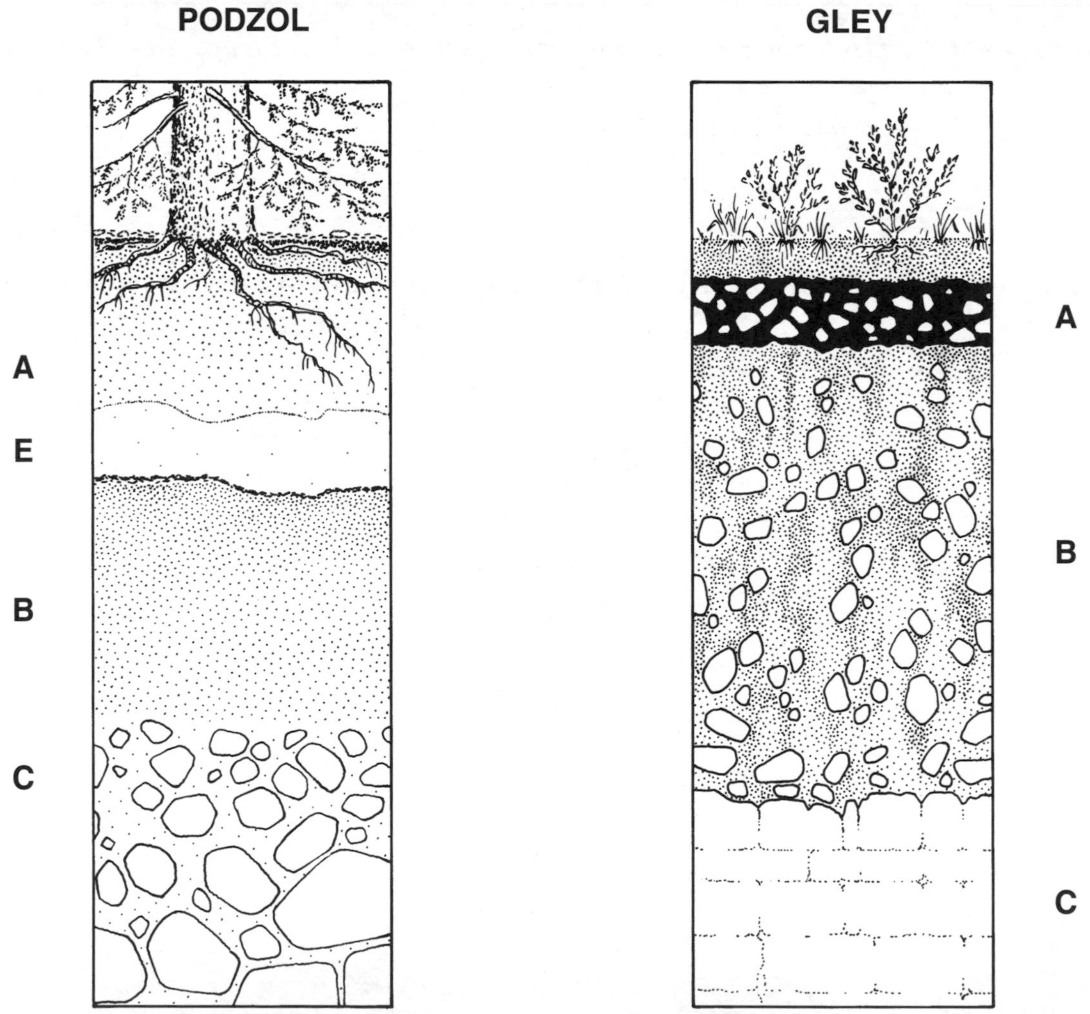

PODZOL

GLEY

(b)   **Explain** how factors such as natural vegetation, soil organisms, climate, relief and drainage have contributed to the formation and characteristics of a **brown earth** soil.   **8**

Marks

**DO NOT ANSWER THIS QUESTION IF YOU HAVE
ALREADY ANSWERED QUESTION 5**

Question 6: Atmosphere

(*a*)   With the aid of an annotated diagram or diagrams, **explain** why there is a surplus of solar energy in the tropical latitudes and a deficit of solar energy towards the poles.   **8**

(*b*)   Study Diagram Q6.

  Describe the possible consequences of global warming throughout the world.   **6**

Diagram Q6: Newspaper Extract

"*Global warming is causing an increase in temperature throughout the World. In Scotland, some areas of Glasgow near the Clyde are in danger of serious flooding—and the risk is only going to get worse because of climate change due to global warming.*"

The Herald, 28 November 2009

[Turn over

Marks

**SECTION C: Answer ONE question from this section,
ie either Question 7 or Question 8.**

Question 7: Rural Geography

(a)   **Describe** the main features of the shifting cultivation farming system.   **6**

(b)   Study Diagram Q7.

Describe and **explain** the advantages and disadvantages of **two** of the **changes** shown in intensive peasant farming.   **8**

Diagram Q7: Recent changes in intensive peasant farming

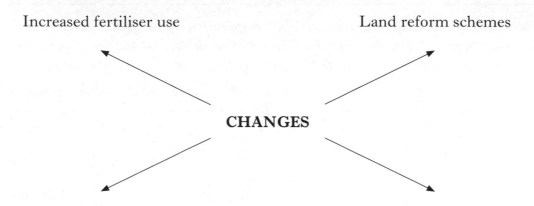

Increased fertiliser use   Land reform schemes

CHANGES

High Yielding Varieties (HYVs) of crops   Mechanisation

Marks

DO NOT ANSWER THIS QUESTION IF YOU HAVE
ALREADY ANSWERED QUESTION 7

Question 8: Urban

(*a*)   For a named Developed World city you have studied, **describe** the changes that have taken place in the Central Business District (CBD).

Your answer should refer to specific named locations within your chosen city.   **8**

(*b*)   Study Diagram Q8.

Describe and **explain** the main urban landscape characteristics of **either** the inner city **or** the outer suburbs.   **6**

Diagram Q8: A summary model of Land Use in a City

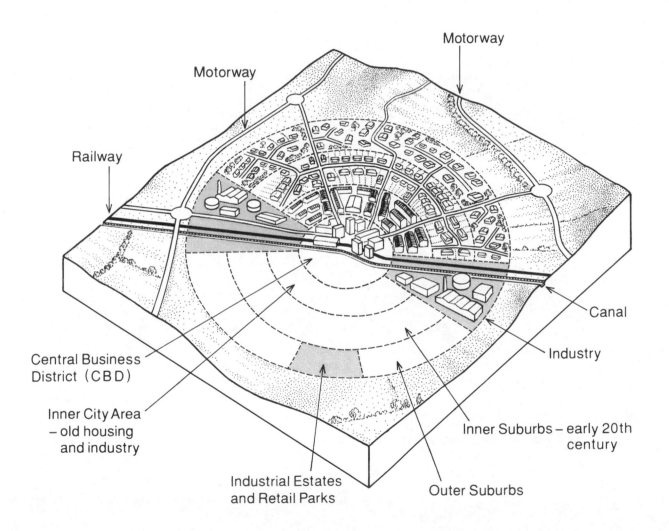

[END OF QUESTION PAPER]

[BLANK PAGE]

X208/12/02

NATIONAL QUALIFICATIONS 2012	TUESDAY, 8 MAY 10.50 AM – 12.05 PM	**GEOGRAPHY** HIGHER Paper 2 Environmental Interactions

Answer any **two** questions.

Write the numbers of the **two** questions you have attempted in the marks grid on the back cover of your answer booklet.

The value attached to each question is shown in the margin.

Credit will be given for appropriate maps and diagrams, and for reference to named examples.

Questions should be answered in sentences.

Note   The reference maps and diagrams in this paper have been printed in black only: no other colours have been used.

Marks

Question 1: Rural Land Resources

(a)   The Yorkshire Dales National Park is an area of Upland Limestone.

With the aid of annotated diagrams, **describe** and **explain** how the main physical features of upland limestone landscapes are formed.

Both surface **and** underground features should be included in your answer.   **20**

(b)   For the Yorkshire Dales National Park, **or** a named upland area you have studied, **describe** how this landscape has provided a variety of socio-economic opportunities.   **10**

(c)   Study Map Q1 and Diagram Q1.

Environmental conflicts such as this windfarm proposal for the Yorkshire Dales may occur in upland landscapes.

With reference to any named upland landscape that you have studied:

(i)   **describe** and **explain** the environmental conflicts;   **10**

(ii) **describe** the measures taken to resolve these environmental conflicts and **comment on** their effectiveness.   **10**

(50)

Map Q1: Location of Yorkshire Dales National Park and Gargrave Windfarm

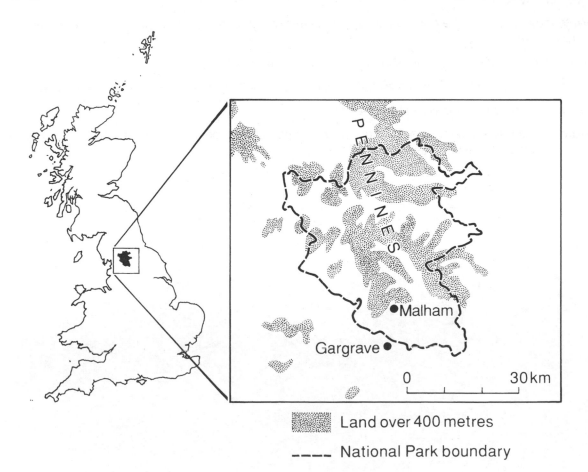

▓▓▓ Land over 400 metres

- - - - National Park boundary

Question 1 – continued

Diagram Q1: Newspaper Extract on Proposals for a wind farm at Gargrave

"*Celebrations as Dales windfarm plan is rejected*.

Hundreds of Yorkshire Dales residents in the Gargrave area were celebrating today over the decision by a government planning inspector to reject plans for a major windfarm which would have dominated the scenery for some 40 miles around. The turbines . . . would have been visible from some of the outstanding beauty spots of the Yorkshire Dales National Park."

Yorkshire Dales Country News, 9 March 2010

[Turn over

Marks

Question 2: Rural Land Degradation

(*a*)   Study Map Q2

Describe the climatic conditions found in Mali and **explain** why such physical conditions may lead to the degradation of rural land.

10

Map Q2: Climatic regions of Mali

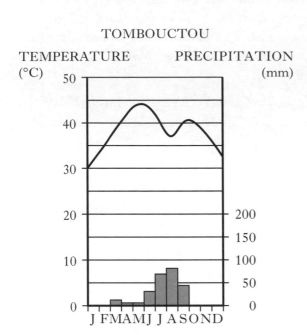

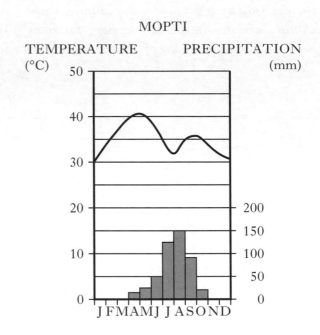

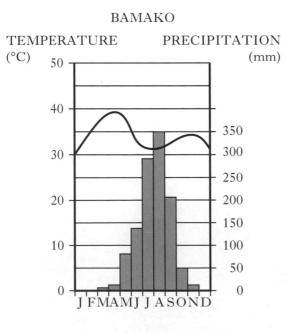

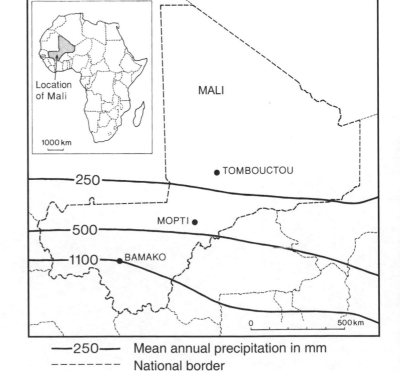

KEY

■ Precipitation (mm)

⌒ Temperature (°C)

Marks

Question 2 – continued

(b)   Study Table Q2A.

Select **two** causes of land degradation from North America **and two** from **either** the Amazon Basin **or** Africa north of the Equator.

Referring to named areas, **explain** how these human activities have contributed to land degradation.

20

Table Q2A: Causes of rural land degradation

North America	Africa north of the Equator OR The Amazon Basin	
Monoculture	Deforestation	Deforestation
Deep ploughing	Overcultivation	Cattle ranching
Farming marginal land	Overgrazing	Mining
Demand for wheat	Population increase	HEP schemes

(c)   Study Table Q2B.

Select **two** soil conservation strategies from North America **and two** from **either** the Amazon Basin **or** Africa north of the Equator.

Referring to named areas:

(i) **describe** your chosen methods and **explain** how they help to conserve soil in rural areas;

(ii) **comment** on the effectiveness of each of your chosen methods.

20

(50)

Table Q2B: Soil conservation strategies

North America	Africa north of the Equator OR The Amazon Basin	
Contour ploughing	Animal fences	Agro-forestry schemes
Diversification	Dams built in gullies	Crop rotation
Shelter belts	Stabilisation of dunes	Return to traditional farming
Strip cropping	"Magic Stones" (Diguettes)	Purchase by conservation groups

[Turn over

Marks

Question 3: River Basin Management

(a) *"The Zambezi river basin extends into 8 southern African countries. It is one of the most heavily dammed rivers in Africa."*

Study Map Q3 and Tables Q3A and Q3B.

Describe and **explain** why there is a need for water management within the Zambezi river basin.   **12**

(b) **Describe** and **explain** the physical and human factors which should be considered when selecting the site for any major dam and its associated reservoir.   **14**

(c) **Describe** and **account for** the social, economic and environmental benefits **and** adverse consequences of a named water control project in Africa **or** Asia **or** North America.   **24**

(50)

Map Q3: Zambezi River Basin

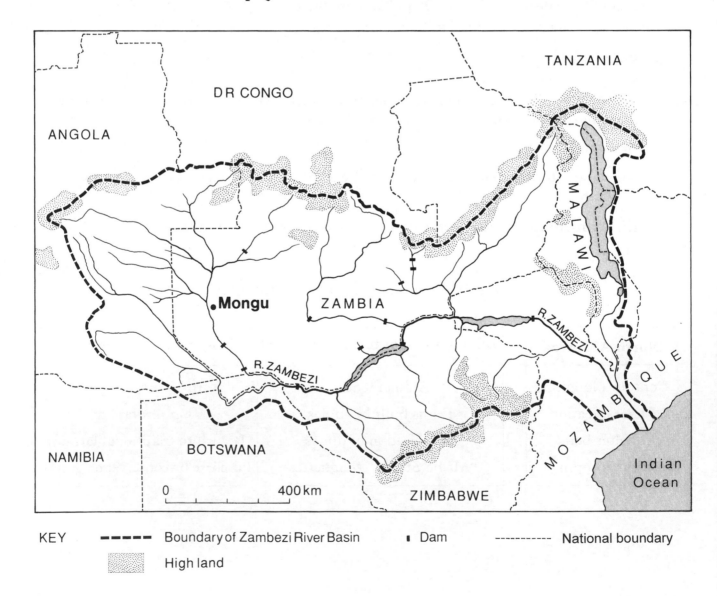

KEY — — — Boundary of Zambezi River Basin   ▪ Dam   - - - - - National boundary

High land

Question 3 – continued

Table Q3A: Zambezi River Basin statistics

Country	% of River basin area	Population growth (%) per annum
Zambia	41	3·1
Angola	18	2·0
Zimbabwe	16	4·3
Mozambique	11	2·4
Malawi	8	2·8
Botswana	3	1·7
Tanzania	2	2·0
Namibia	1	0·9

Table Q3B: Climate figures for Mongu

	J	F	M	A	M	J	J	A	S	O	N	D
Max temperature (°C)	29	29	29	30	28	27	27	30	33	34	31	29
Precipitation (mm)	210	185	140	45	5	1	0	2	2	35	105	195

[Turn over

Marks

Question 4: Urban Change and its Management

(*a*)   Study Map Q4A and Map Q4B.

Describe and account for the distribution of major settlements in either Australia or any other Developed Country that you have studied.   8

(*b*)   Study Map Q4C on *Page ten*.

With reference to Melbourne, or any named city you have studied in a Developed Country, explain ways in which its site and situation contributed to its growth.   8

(*c*)   Study Map Q4D on *Page ten*.

Referring to Melbourne or any named city you have studied in a Developed Country:

  (i)   explain the problems caused by urban sprawl;

  (ii) suggest ways in which these problems may be resolved.   12

(*d*)   "*Lagos is one of the world's mega-cities—a crime-ridden, seething mass of some 15 million people crammed into the steamy lagoons of Southwest Nigeria. Two out of three Lagos residents live in a slum. The government estimates that Lagos will have expanded to 25 million residents by 2015 to be the third largest city in the world.*"

For Lagos or any named city you have studied in a Developing Country:

  (i)   explain why your chosen city has grown so rapidly;   10

  (ii) describe the socio-economic and environmental problems which have resulted from such rapid growth.   12

(50)

Question 4 – continued

Map Q4A: Largest cities in Australia

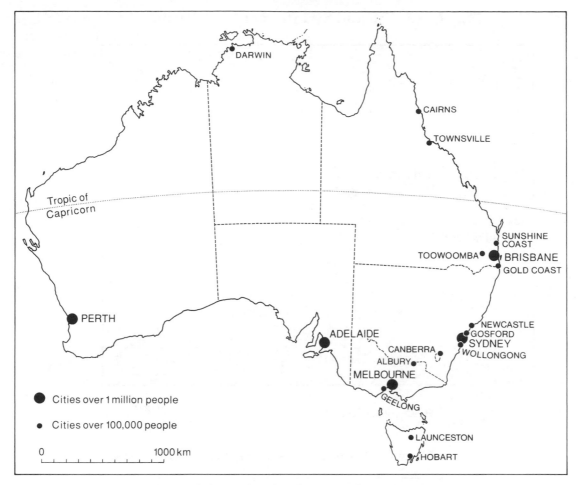

Map Q4B: Physical Map of Australia

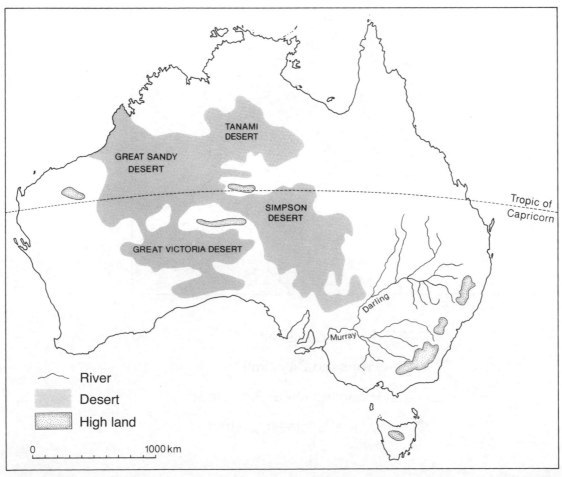

Question 4 – continued

Map Q4C: Site and Situation of Melbourne

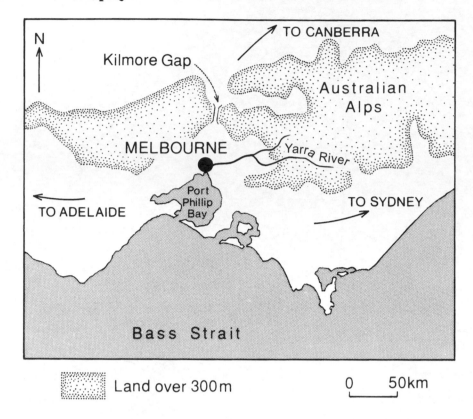

Land over 300 m   0   50km

Map Q4D: New Melbourne Urban Growth Boundary

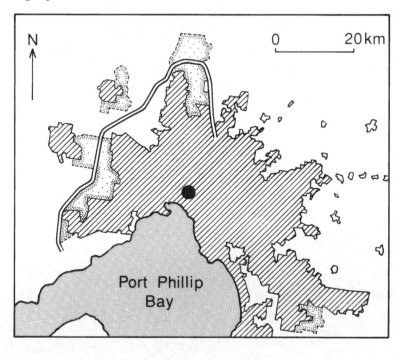

/////// Current City Limit

.... Proposed City Limit

====== Proposed Outer Ring Road

● Central Business District

[Turn over for Question 5 on *Page twelve*

Marks

Question 5: European Regional Inequalities

(a)   Study Maps Q5A and Q5B.

Convergence Region Funding aims to raise the standard of living in the poorest EU countries and replaced Objective 1 Funding.   This funding covers regions whose GDP per capita is below 75% of the EU average and aims at accelerating their economic development.

(i) **Describe** the changing distribution of areas towards which EU funding is directed.   **8**

(ii) **Explain** how EU initiatives such as Convergence Region Funding might improve the development of less prosperous areas of the European Union.   **10**

Map Q5A: **2003–2007 EU Objective 1 Funding**	**Map Q5B:** **2007–2012 Convergence Region Funding**

Regions eligible under Objective 1

Convergence Regions receiving most financial aid

(b)   Study Table Q5

(i) **Describe** how the data shows a pattern of inequality across the European Union member states.   **10**

(ii) **Suggest** both physical **and** human reasons for the variation in prosperity found across the 27 European Union member states.   **14**

(c)   *"Many European countries suffer from regional inequalities **within** them."*

For any named European Union country you have studied **describe** the steps taken by the national government to reduce inequalities and **comment** on their effectiveness.   **8**

(50)

Question 5 – continued

Table Q5: European Union Statistics, 2011

Country (Year of membership)	HDI (ranked)*	GDP per capita (Euros)	% of population aged 15–64 in employment	% Internet users
Belgium (1957)	10	28 200	62	78
France (1957)	4	26 300	64	69
Germany (1957)	12	26 900	71	79
Italy (1957)	11	24 300	58	52
Luxembourg (1957)	5	65 700	65	85
Netherlands (1957)	2	30 700	77	89
Denmark (1973)	7	29 600	76	86
Ireland (1973)	1	34 200	62	66
UK (1973)	12	27 800	70	83
Greece (1981)	14	22 900	61	46
Portugal (1986)	17	17 500	66	48
Spain (1986)	7	24 700	60	63
Finland (1995)	6	27 500	69	85
Sweden (1995)	3	29 300	72	93
Austria (1995)	7	30 000	72	75
Cyprus (2004)	16	21 603	70	39
Czech Republic (2004)	18	18 500	67	66
Estonia (2004)	20	16 100	70	75
Hungary (2004)	23	15 300	55	62
Latvia (2004)	25	12 600	61	68
Lithuania (2004)	24	13 200	60	59
Malta (2004)	19	18 100	55	59
Poland (2004)	21	12 300	59	58
Slovakia (2004)	21	15 000	60	74
Slovenia (2004)	15	20 700	68	65
Bulgaria (2007)	26	8600	64	48
Romania (2007)	27	9100	59	36

HDI Human Development Index (combined indicator which includes a measure of wealth, health and education in a country)

* Ranking 1–27 with 1 best and 27 worst

Marks

Question 6: Development and Health

(a) Infant Mortality Rate per 1000 live births is a social indicator of development.

Name **one** other social indicator and **one** economic indicator of development and **explain** how they show a country's level of development.   **8**

(b) Referring to named **developing countries** that you have studied, **suggest reasons** why there is such a wide range in levels of development **between** developing countries.   **12**

(c) Study Table Q6 and Map Q6.

Many African countries have been trying to eliminate water-related diseases like malaria, cholera and bilharzia/schistosomiasis.

For **one** of the above diseases:

(i) **describe** the physical **and** human factors which put people at risk of contracting the disease;   **8**

(ii) **describe** the measures that can be taken to combat the disease.   **12**

(d) Study Diagram Q6.

Many Developing Countries are attempting to reduce the death rate of children under 5 by implementing Primary Health Care strategies.

Describe some specific Primary Health Care strategies and explain why these strategies are suited to people living in developing countries.   **10**

(50)

Table Q6: Statistics on Malaria

- 3·3 billion people in 109 countries are at risk from malaria
- 247 million annual cases of malaria
- 850 000 people die from malaria each year
- 91% of deaths caused by malaria are in Africa
- 85% of deaths caused by malaria are of children aged under 5

Extract from article "Why can't we rid the world of malaria?"

Daily Telegraph, 8 July 2010

Question 6 – continued

Map Q6: Areas in Africa affected by Malaria

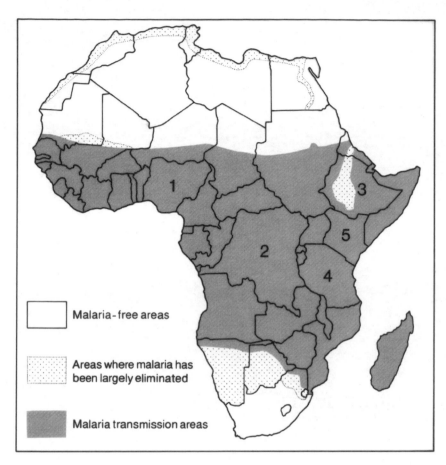

Most infected countries

1   Nigeria (57·5 million)
2   DR Congo (23·6 million)
3   Ethiopia (12·4 million)
4   Tanzania (11·5 million)
5   Kenya (11·3 million)

In brackets are the number of people infected.

Malaria-free areas

Areas where malaria has been largely eliminated

Malaria transmission areas

Diagram Q6: Major causes of deaths of children under 5 in the Developing World, 2008

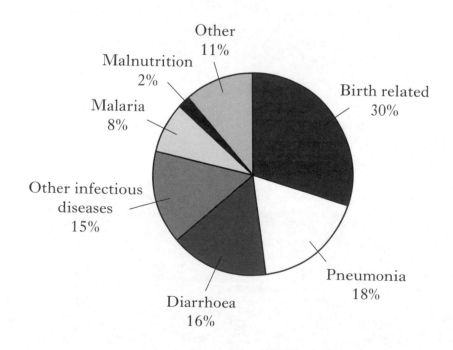

Other
11%

Malnutrition
2%

Malaria
8%

Other infectious
diseases
15%

Diarrhoea
16%

Pneumonia
18%

Birth related
30%

[END OF QUESTION PAPER]

[BLANK PAGE]

[BLANK PAGE]

X208/12/01

NATIONAL QUALIFICATIONS 2013	THURSDAY, 30 MAY 9.00 AM – 10.30 AM	GEOGRAPHY HIGHER Paper 1 Physical and Human Environments

Six questions should be attempted, namely:

all four questions in **Section A** (Questions 1, 2, 3 and 4);

one question from **Section B** (Question 5 **or** Question 6);

one question from **Section C** (Question 7 **or** Question 8).

Write the numbers of the **six** questions you have attempted in the marks grid on the back cover of your answer booklet.

The value attached to each question is shown in the margin.

Credit will be given for appropriate maps and diagrams, and for reference to named examples.

Questions should be answered in sentences.

Note The reference maps and diagrams in this paper have been printed in black only: no other colours have been used.

Extract No 2006/OL20

1:25 000 Scale
Explorer Series

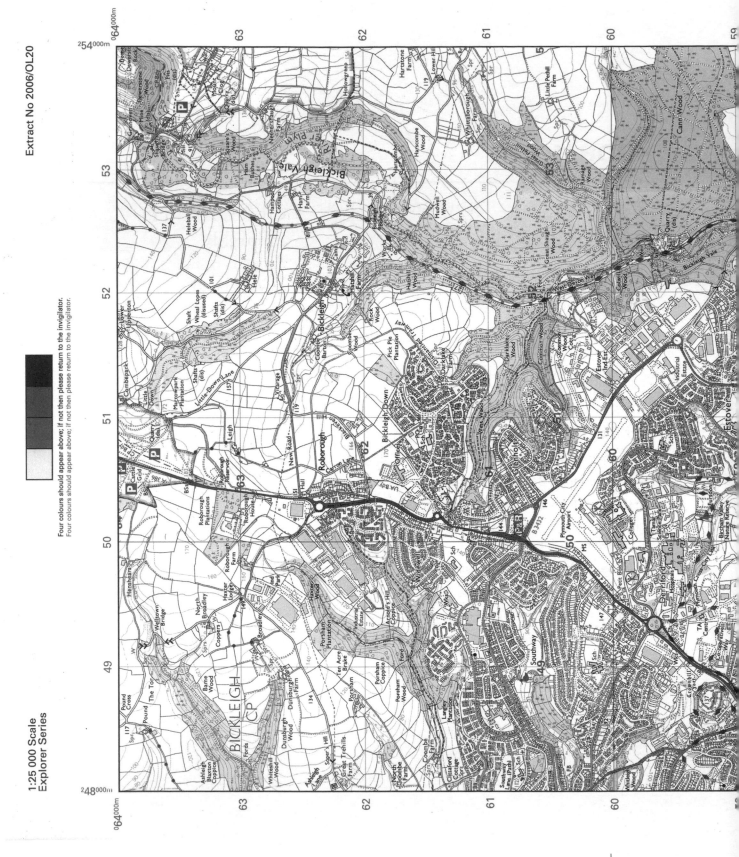

Magnetic North
Grid North
True North

Diagrammatic
only

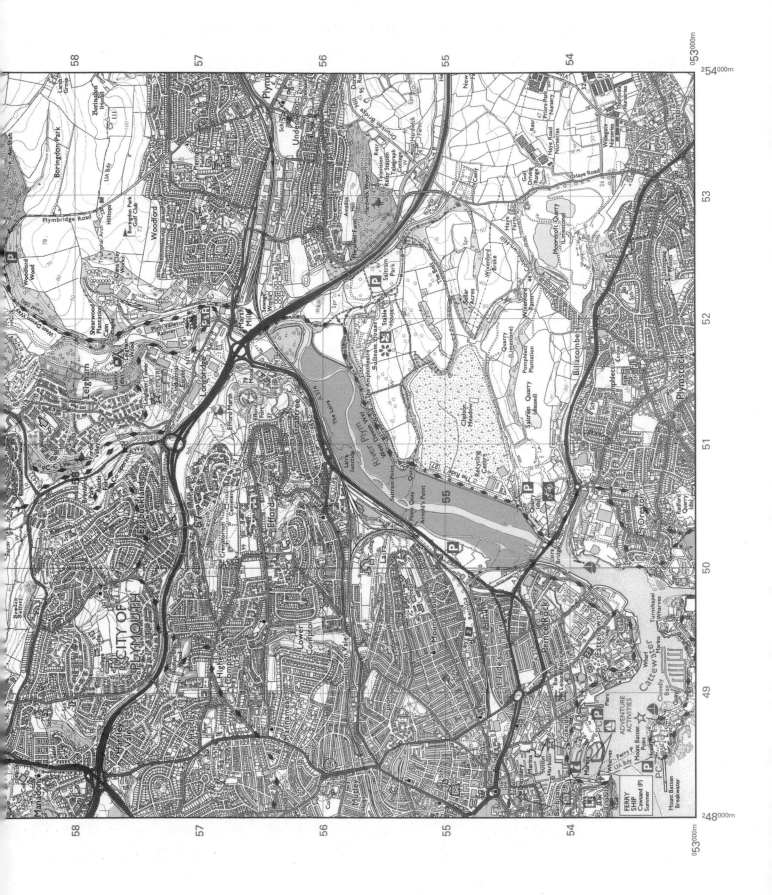

Scale 1: 25 000

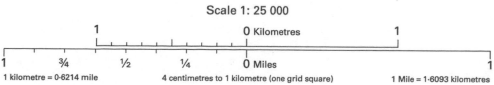

1 kilometre = 0·6214 mile   4 centimetres to 1 kilometre (one grid square)   1 Mile = 1·6093 kilometres

Marks

SECTION A: Answer ALL four questions from this section.

Question 1: Lithosphere

Study Diagram Q1 which shows a typical surface landscape and cave system in the Yorkshire Dales, an area with Carboniferous Limestone features.

(a)   Select **one** surface and **one** underground feature from the lists below.

Describe and **explain** the formation of both features. You may use an annotated diagram or diagrams in your answer.

Surface Features	Underground Features
Limestone pavement	Stalactites and stalagmites
Swallow hole	Cave/Cavern

12

(b)   Scree slopes are often found at the bottom of cliffs or scars typical of Carboniferous Limestone landscapes. **Explain** the processes involved in their formation.

6

Diagram Q1: A Typical Carboniferous Landscape

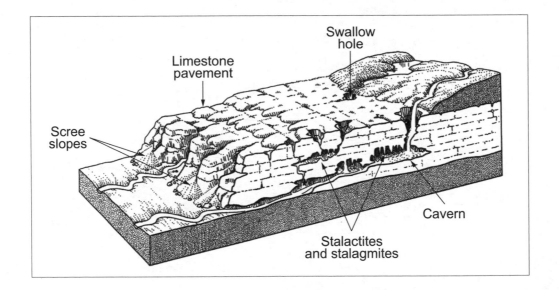

Marks

Question 2: Atmosphere

(*a*)   Study Diagram Q2A and Q2B.

Describe and **explain** why the Earth's surface absorbs only 50% of the solar energy received at the edge of the atmosphere. You should refer to both conditions in the Earth's atmosphere and at the Earth's surface.   **8**

(*b*)   There has been an increase in the average global temperature in the last 150 years.

Describe and **explain** the **human** factors affecting global warming.   **10**

Diagram Q2A: Earth/Atmosphere Energy Exchange

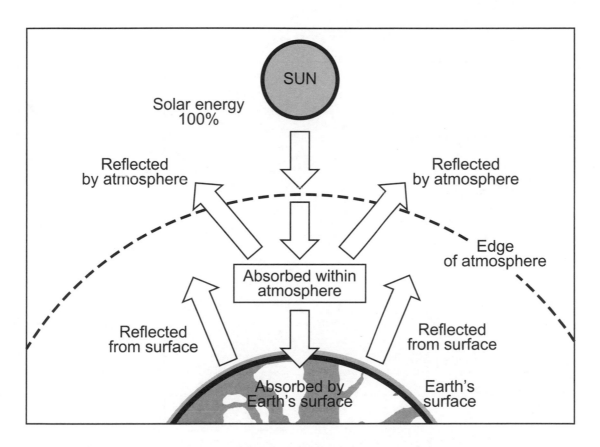

Diagram Q2B: Proportion of solar energy absorbed/reflected

Absorbed by Earth's surface	Reflected by atmosphere	Absorbed within atmosphere	RS

0   50   100%

RS = Reflected from surface

[Turn over

Marks

Question 3: Urban

(a) Study OS Map Extract number 2006/OL20: Plymouth (*separate item*), **and** Map Q3.

Using map evidence, **describe** the residential environments of Area A and Area B.
Suggest reasons for the differences.   **12**

(b) **Suggest** the **impact** that an out of town shopping centre may have had on the traditional Central Business District (CBD) of Plymouth or any other named city you have studied in a developed country.   **6**

Map Q3 : Location of residential areas in Plymouth

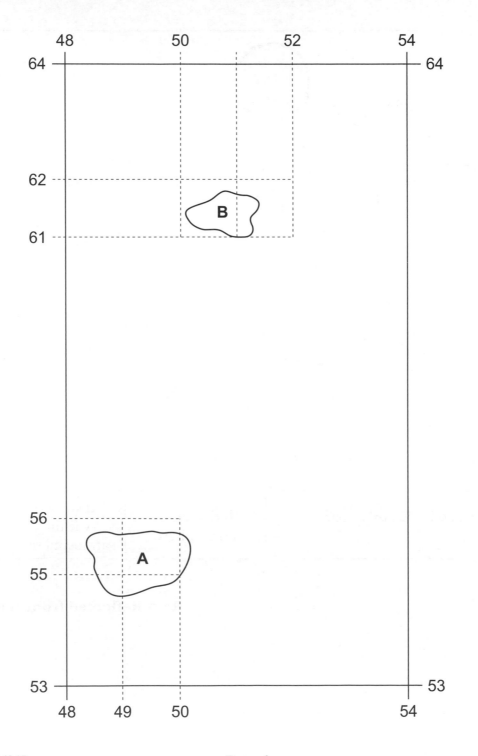

Marks

Question 4: Rural

Study Diagram Q4.

Choose **one** of these farming systems.

Referring to a named area where your chosen system is carried out:

(i) **explain** the ways in which the diagram reflects the main features of your chosen
system;   **8**

(ii) **describe** the recent changes in farming practices that have taken place and
discuss the impact of these changes on the people and their environment.   **10**

Diagram Q4: Farming systems

Intensive Peasant Farming   **Commercial Arable Farming**

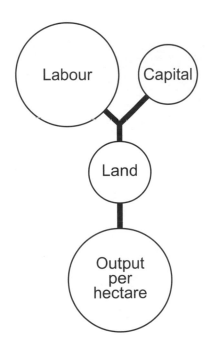 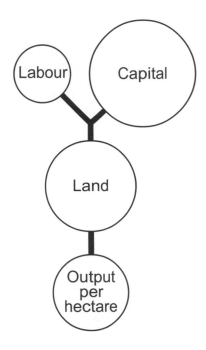

[Turn over

Marks

SECTION B: Answer ONE question from this section, ie either Question 5 or Question 6.

Question 5: Hydrosphere

Study OS Map Extract number 2006/OL20: Plymouth (*separate item*).

(*a*)　Using appropriate grid references, **describe** the **physical** characteristics of the River Plym **and** its valley from Bickleigh Bridge (GR 527618) to Laira Bridge (GR 501542).

8

(*b*)　**Explain**, with the aid of an annotated diagram or diagrams, how a meander is formed.

6

Marks

DO NOT ANSWER THIS QUESTION IF YOU HAVE
ALREADY ANSWERED QUESTION 5

Question 6: Biosphere

(*a*)   Study Diagram Q6 which shows a coastal sand dune area.

Describe and give reasons for the plant types likely to be present at one of the locations A, B or C. Named plant species should be included.   **6**

(*b*)   Draw and fully annotate a soil profile of a brown earth soil to show its main characteristics (including horizons, colour, texture, soil biota and drainage) and associated vegetation.   **8**

Diagram Q6: Transect across sand dune coastline

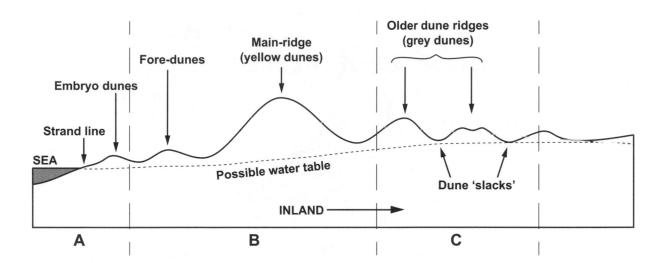

[Turn over

Marks

SECTION C: Answer ONE question from this section, ie either Question 7 or Question 8.

Question 7: Population

Study Diagram Q7A

(a) **Describe** and **explain** the population structure of Malawi in 2010.   **8**

Study Diagram Q7B.

(b) **Discuss** the possible consequences of the 2050 population structure for the future economy of Malawi and the welfare of its citizens.   **6**

Diagram Q7A: Population Pyramid for Malawi, 2010

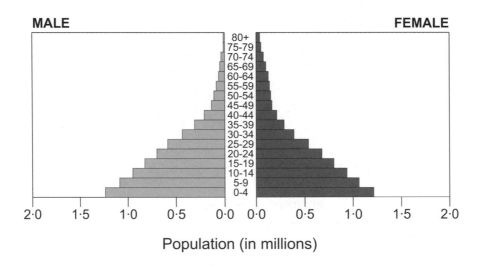

Diagram Q7B: Projected Population Pyramid for Malawi, 2050

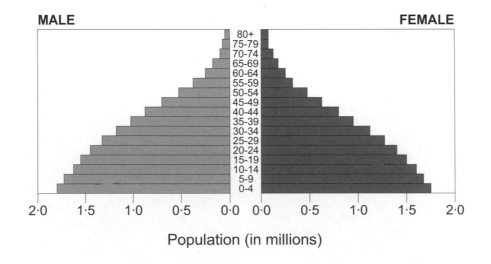

Marks

DO NOT ANSWER THIS QUESTION IF YOU HAVE ALREADY ANSWERED QUESTION 7

Question 8: Industrial Geography

Study photographs Q8A, Q8B and Q8C.

"New" industry is often located in industrial estates, business parks and science parks.

Referring to a **named** industrial concentration in the European Union that you have studied:

(i) **describe** and **explain** the main characteristics of a typical new industrial landscape;   **7**

(ii) **describe** ways in which the European Union **and** national governments have helped to attract new industries to your chosen area.   **7**

Photograph Q8A   **Photograph Q8B**

Photograph Q8C

[END OF QUESTION PAPER]

[BLANK PAGE]

X208/12/02

NATIONAL QUALIFICATIONS 2013	THURSDAY, 30 MAY 10.50 AM – 12.05 PM	GEOGRAPHY HIGHER Paper 2 Environmental Interactions

Answer any **two** questions.

Write the numbers of the **two** questions you have attempted in the marks grid on the back cover of your answer booklet.

The value attached to each question is shown in the margin.

Credit will be given for appropriate maps and diagrams, and for reference to named examples.

Questions should be answered in sentences.

Note   The reference maps and diagrams in this paper have been printed in black only: no other colours have been used.

Marks

Question 1: Rural Land Resources

(*a*) **Describe** and **explain**, with the aid of annotated diagrams, the formation of the main features of glaciation in the Cairngorms National Park or any other glaciated upland area you have studied.

20

(*b*) Study Diagram Q1.

With reference to the Cairngorms National Park or any other upland area you have studied, explain the social **and** economic opportunities created by the landscape.

8

(*c*) Referring to named examples within the Cairngorms or any other upland or coastal area you have studied:

(i) **describe** and **explain** the environmental conflicts that have occurred;

11

(ii) **describe** the solutions to these environmental conflicts commenting on their effectiveness.

11

(50)

Diagram Q1: The Cairngorms Mountain Range

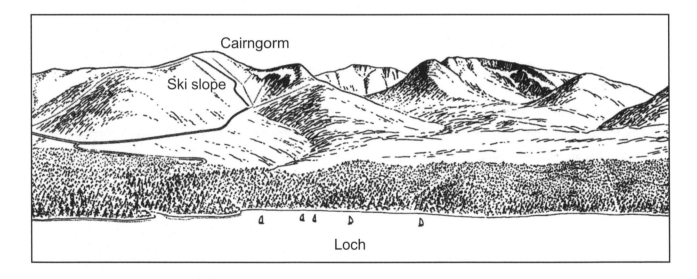

Marks

Question 2: Rural Land Degradation

(a)   Study Diagram Q2.

Describe and **explain** the processes of soil erosion by wind.   6

Diagram Q2: Selected processes of wind erosion

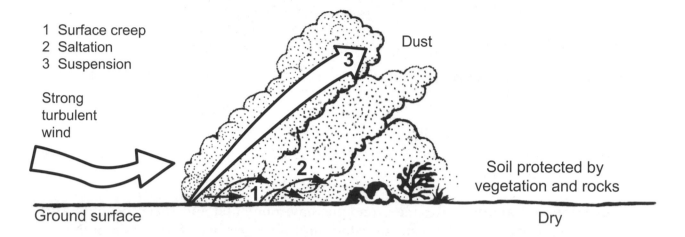

1 Surface creep
2 Saltation
3 Suspension

Strong
turbulent
wind

Dust

Soil protected by
vegetation and rocks

Ground surface

Dry

(b)   **Describe** and **explain** how human activities, including inappropriate farming techniques, have caused land degradation in North America.   14

(c)   Referring to named locations in **either** Africa north of the Equator **or** the Amazon Basin, **describe** the impact of land degradation on the people, economy and the environment.   10

(d)   For named areas in North America **and** Africa north of the Equator **or** the Amazon Basin:

   (i)   **describe** and **explain** soil conservation strategies that have reduced land degradation;

   (ii)   **comment** on the effectiveness of these strategies.   20

                    (50)

[Turn over

Marks

Question 3: River Basin Management

(*a*)   Study Maps Q3A, Q3B and Q3C.

   For North America, Africa or Asia, **describe** and **explain** the general distribution of river basins.

9

(*b*)   "*The Mississippi river basin extends into 31 states of the USA as well as into southern Canada. It is the third largest river basin in the world*".

   Study Maps Q3A, Q3D and Diagram Q3.

   Describe and **explain** why there is a need for water management within the Mississippi River Basin.

10

(*c*)   For the Mississippi River Basin or any other river basin management project in North America **or** Africa **or** Asia, **explain** the **political** problems that may have resulted from the project.

7

(*d*)   **Describe** and **account for** the economic, environmental and social benefits **and** adverse consequences of a named water control project in Africa, Asia or North America.

24

(50)

Question 3 – continued

Map Q3A:
Major river basins of North America

Map Q3B:
Major river basins of Africa

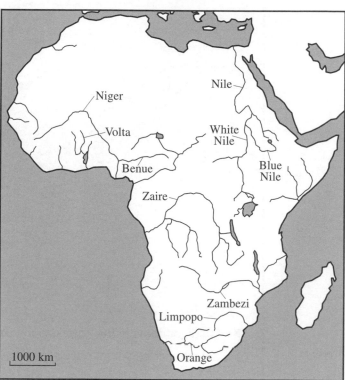

Map Q3C: Major river basins of Asia

Question 3 – continued

Map Q3D: Mississippi River Basin

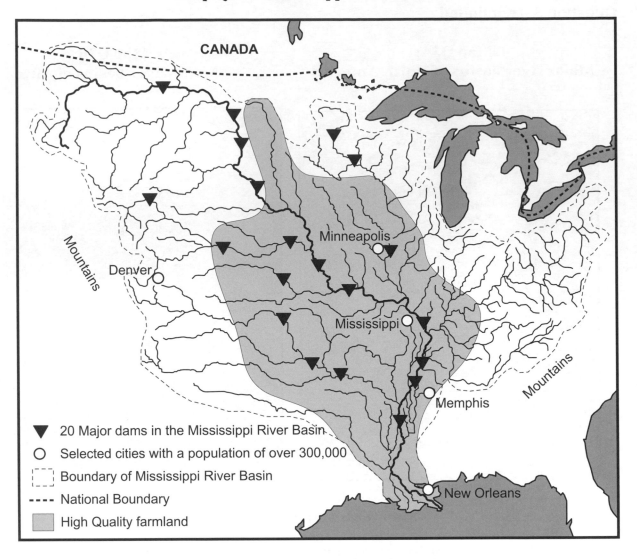

▼ 20 Major dams in the Mississippi River Basin

○ Selected cities with a population of over 300,000

⬚ Boundary of Mississippi River Basin

- - - National Boundary

▨ High Quality farmland

Diagram Q3: Climate Graphs

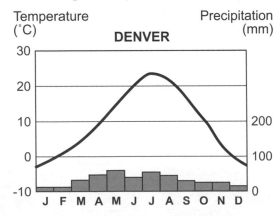

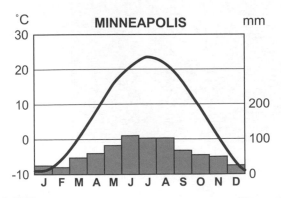

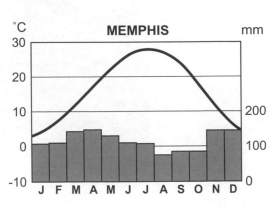

[Turn over for Question 4 on *Page eight*

Marks

Question 4: Urban Change and its Management

(a)   Study Map Q4A.

Describe and **account** for the projected distribution of the world's largest urban areas.

14

(b)   Study Map Q4B.

Referring to Tokyo or any other named city that you have studied in the Developed World:

    (i)   **outline** the problems caused by urban sprawl.

    (ii)   **explain** the ways in which the city has tried to resolve this problem and **comment** on their effectiveness.

18

(c)   Study Diagram Q4.

With reference to a **named** city that you have studied in the Developing World:

    (i)   **describe** the social, economic and environmental problems found in shanty town areas;

12

    (ii)   **describe** the methods the residents and local authorities have used to tackle these problems.

6

(50)

Map Q4A: Twelve largest urban areas in the world 2015 (projection)

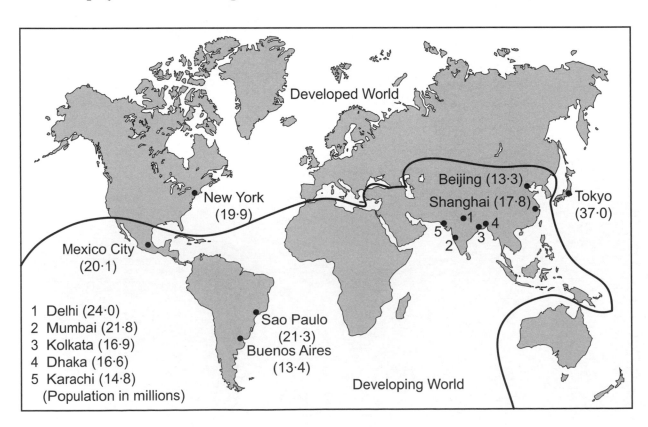

Question 4 – continued

Map Q4B: Urban Growth of Tokyo 1945–2013

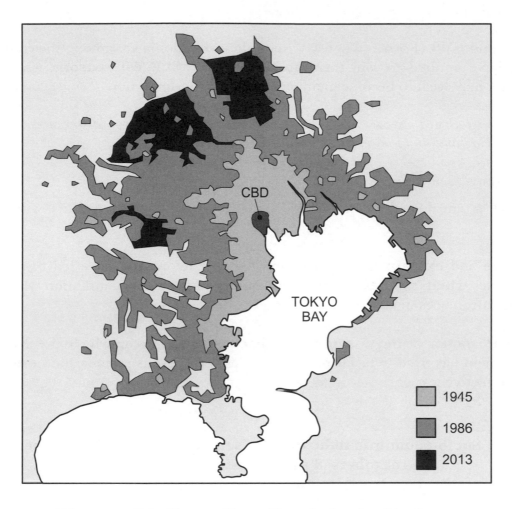

	1945
	1986
	2013

Diagram Q4: Shanty Town Population by Continent

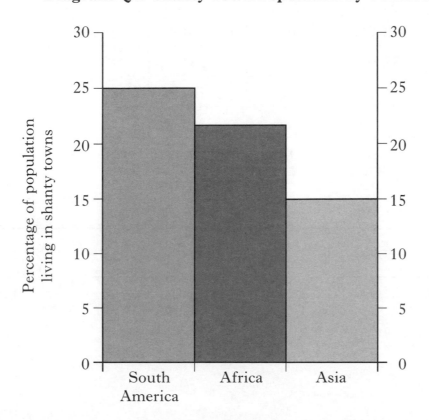

Marks

Question 5: European Regional Inequalities

(a) Study Table Q5A.

Turkey and FYR (Former Yugoslav Republic) Macedonia are among the countries which have applied to join the European Union. **Suggest reasons** why these countries may wish to become members of the European Union. **12**

(b) *"The North-South divide refers to the economic and cultural differences between southern England and the rest of the United Kingdom."*

Study Map Q5 and Table Q5B.

To what extent does the data provide evidence of regional inequalities within the UK? **12**

(c) **Describe** and **explain** the physical and human factors that have led to regional inequalities within the UK or any other country of the European Union which has marked differences in economic development between regions. **15**

(d) For your chosen country in part (c), **discuss** the ways in which the National Government has tried to tackle problems in less prosperous regions and **comment** on the effectiveness of these strategies. **11**

(50)

Table Q5A: Socio-economic indicators for selected current and prospective members of the European Union

Country	Year of joining EU	GDP per capita 2010 (PPP*)	Industrial Sector (%) 2010			Unemployment (%) 2010
			Primary	Secondary	Tertiary	
Belgium	1957	37,800	2	25	73	8·5
UK	1973	34,800	2	18	80	7·9
Portugal	1986	23,200	12	20	68	10·7
Bulgaria	2007	13,500	6	30	64	9·2
FYR Macedonia	–	9,700	20	22	58	31·7
Turkey	–	12,300	29	25	46	12·4

PPP* = Purchasing Power Parity

Question 5 – continued Map Q5: UK statistical regions

Table Q5B: Selected indicators of development for UK regions

	Gross disposable household income 2010 (UK average = 100)	Average house prices 2011 × £1,000	Projected population change (%)	
			2009–2014	2009–2019
Scotland	94	146	0·4	0·6
Northern Ireland	85	144	3·1	4·9
Wales	87	146	2·1	4·4
NW England	91	151	2·3	5·0
NE England	85	143	1·1	2·4
West Midlands	92	167	2·5	5·3
Yorks & Humber	91	150	4·5	8·9
East Midlands	94	156	4·8	9·2
East England	107	196	4·5	9·0
SE England	115	273	4·1	9·4
London	120	437	5·5	9·9
SW England	99	223	4·7	8·0
UK average	**100**	**233**	**3·6**	**7·0**

[Turn over for Question 6 on *Page twelve*

Marks

Question 6: Development and Health

(a) "Number of people per doctor" is an example of a social indicator of development.

Name fully **two** other social indicators and **two** economic indicators which might identify different levels of development.   8

(b) Using named examples, **suggest reasons** for the wide variations in development which exist **between** Developing Countries.   12

(c) Study Map Q6.

Malaria, cholera and bilharzia/schistosomiasis are water related diseases which remain the biggest causes of death in Developing Countries.

Select **one** of the diseases above.

(i) **Describe** the physical **and** human factors which put people at risk of contracting the disease.   8

(ii) **Describe** the measures that can be taken to combat the disease and **explain** the varying effectiveness of these measures.   17

(iii) **Explain** the benefits to a Developing Country of controlling the disease.   5

(50)

Map Q6: Countries affected by Malaria

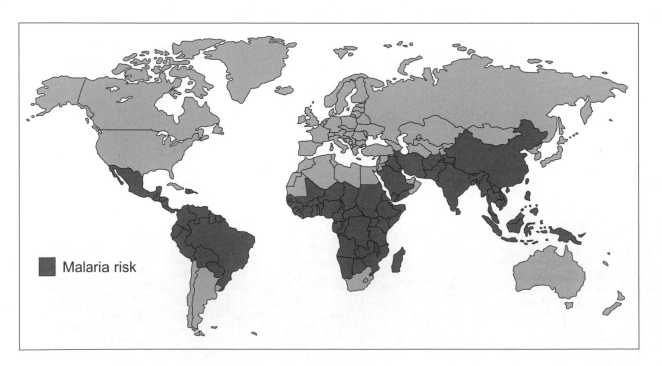

Malaria risk

[END OF QUESTION PAPER]

[BLANK PAGE]

Acknowledgements

Permission has been sought from all relevant copyright holders and Hodder Gibson is grateful for the use of the following:

A photograph reproduced by permission of Buchanan Galleries (2009 Paper 1 page 10);

A photograph reproduced by permission of Braehead shopping centre. (2009 Paper 1 page 10);

An extract from 'Relief for York, for now,' taken from BBC News, Saturday 4 November 2000 © BBC (2010 Paper 1 page 6);

Two extracts adapted from 'Trump golf resort would destroy jewel in crown of bare dunes' by Graeme Smith, taken from The Herald 19/6/08. © Herald & Times Group (2010 Paper 2 page 2);

An extract from http://www.rspb.org.uk/ourwork/conservation/sites/scotland/menie.asp © RSPB (2010 Paper 2 page 2);

An extract © Press Association, December 2009 (2012 Paper 1 page 5);

An extract taken from 'The Herald', 28 November 2009 © Herald & Times Group (2012 Paper 1 page 7);

An extract taken from 'Yorkshire Dales Country News', 9 March 2010 © Dales.net (2012 Paper 2 page 3);

An extract taken from the article 'Why can't we rid the world of malaria?' taken from 'Daily Telegraph' © Telegraph Media Group Limited 8 July 2010 (2012 Paper 2 page 14);

The Eurocentral logo © Eurocentral Partnership Limited (2013 Paper 1 page 9);

Ordnance Survey maps © Crown Copyright 2013. Ordance survey 100047450.

HIGHER | ANSWER SECTION

SQA HIGHER GEOGRAPHY
2009–2013

GEOGRAPHY HIGHER
PHYSICAL AND HUMAN ENVIRONMENTS
2009

SECTION A

Question 1 – Hydrosphere

(a) Descriptions could include:
  • meanders, waterfalls (eg 979664 or 999634), tributaries/confluences, braiding/eyots/islands (982653), river cliffs (Loup Scar at 030618).
  • variations in the width of the valley eg broad, flat flood plain – approx 700 metres wide – in 9768 whereas in other sections such as in square 0162 the river is flowing in more of a gorge.
  • references to the height of the land, steepness of the valley sides, direction of flow.
  • accept direction S/SE.
  • reference to speed only if explained or names.

(b) Answers should be based on the concept of differential erosion. The following points may be made:
  • harder rock overlying softer rock.
  • softer rock is eroded more easily by the force of running water.
  • eventually, the softer rock is worn away.
  • this causes undercutting so there is nothing to support the harder rock above which collapses.
  • some of this shattered rock will be swirled around by the river (especially at times of spate) and helps to excavate a deep plunge pool at the base of the waterfall.
  • this process is repeated over a long period of time causing the waterfall to gradually retreat upstream usually leaving a steep-sided gorge.

Question 2 – Biosphere

(a) A podzol soil profile should be drawn and annotations could include:

Associated vegetation is coniferous forest or heather moorland.

Thin black humus layer divided between layers of leaf litter (L), fermentation (F) and mor humus (H) with a pH of 3.5-4. Plants have shallow, spreading roots.

Ash-grey upper A horizon with sandy texture.

Zone of eluviation of humus, Fe and Al minerals and clay. Well-defined horizons – few soil biota to mix soil due to cold climate.

Iron pan develops in upper B horizon, impeding drainage and causing waterlogging.

Zone of illuviation with accumulation of clay, and Fe and Al oxides.

B horizon is reddish-brown with denser texture. Precipitation exceeds evaporation, giving downward leaching.

C horizon is parent material, generally weathered rock or glacial or fluvio-glacial material.

(b) The following features could be included for a brown earth soil:
  • deciduous forest vegetation provides deep leaf litter, which is broken down rapidly in mild/warm climate.
  • soil colour varies from black humus to dark brown in A horizon to lighter brown in B horizon where humus content is less obvious.

  • soil biota break down leaf litter producing a mildly acidic mull humus. They also ensure the mixing of the soil, aerating it and preventing the formation of distinct layers within the soil.
  • texture is loamy and well-aerated in A horizon but lighter in the B horizon.
  • precipitation slightly exceeds evaporation, giving downward leaching of the most soluble minerals and the possibility of an iron pan forming, impeding drainage.
  • trees have roots which penetrate deep into the soil, ensuring the recycling of minerals back to the vegetation.

Question 3 – Rural Geography

(a) The following points should be developed:
  • Bangladesh is an ELDC with a very low GDP per capita.
  • a high percentage of Bangladesh's low GDP is derived from its intensive peasant farming.
  • a very high percentage of Bangladesh's working population are intensive peasant farmers.
  • due to the intensive nature of farming in Bangladesh the amount of fertiliser used is high, although still less than Canada.
  • Canada has a highly mechanised form of commercial arable farming.

(b) (i) Descriptions might include:
    • improved irrigation.
    • increased farm sizes and larger fields.
    • increased use of fertilizer.
    • increased mechanisation.
    • 'green revolution' type changes eg development of hybrid seeds.
    • use of appropriate technology.
    • increasing export of farming produce.

   (ii) The impact of these changes might include:
    • greater amount of food has reduced malnutrition and starvation.
    • surplus crop may be sold, improving quality of life.
    • increased mechanisation may lead to reduction in farm labour.
    • migration of farm workers to urban areas and impact on demography of rural areas.
    • consolidation of farms may also lead to larger fields, increased mechanisation and drift to cities.
    • improved infrastructure including increased electrification and better roads improving access to markets.
    • larger, more effective irrigation schemes and drainage systems.
    • increased use of insecticides, pesticides and fertilisers may impact on the environment and humans.

Question 4 – Industrial Geography

(i) The following points might be included in a description/explanation of the major characteristics of a *"new"* *industrial landscape*.
  • Lower, smaller, modern buildings – mostly single storey and often with large windows to allow in plenty of light.
  • Buildings are well planned/spaced out with trees and grassy areas and even ornamental lakes/ponds included in the layout to provide a more attractive working environment and create a favourable image to prospective investors/clients.

- These areas are usually located on purpose-built industrial estates or Science/Business Parks commonly on Greenfield sites on the edge of towns/cities where land is relatively cheaper and there is room for car parking and for future expansion.
- They are close to major roads such as dual carriageways or motorways for ease of transport of the finished products to markets/ports, for bringing in raw materials/components/sub assemblies and for the convenience of to-day's more mobile, car-owning workforce.
- There is an absence of slag heaps/coal bings, factory chimneys, railway sidings etc usually associated with older, 'smokestack' industrial areas.
- There is a tendency for similar sorts of industries/firms in similar looking buildings to be located on the same site to benefit from an exchange of ideas and information. Many of these businesses are connected with information, high technology and electronics industries and will have direct links with universities (often situated close by) for research and development purposes and to remain successful and competitive.

(ii) Answers will, of course, depend on the industrial concentration chosen.

National government measures may include:

- incentives such as capital allowances, retraining grants, provision of purpose-built premises/advance factories, rent-free periods, tax relief on new machinery, reduced interest rates…
- specific financial assistance to old industrial areas such as coalfields in UK.
- the creation of Enterprise Zones, Development Areas, Urban Development Corporations. Regional Selective Assistance (Scotland/Wales). Selective Finance for Investment (SFFI – England).
- setting up Manpower Creation Schemes (MCS) and Youth Training Schemes (YTS).
- decentralising/relocating government offices/departments (eg DVLA to Swansea or National Savings to Glasgow).
- Improving communications/accessibility/existing infrastructure.

EU assistance could include mention of such agencies as ERDF (European Regional Development Fund), EIB (European Investment Bank), ESF (European Social Fund), etc and their associated benefits and/or more general references to EU funding/regeneration projects.

SECTION B

Question 5 – Atmosphere

(a) Description and Explanation might include:

- currents follow loops or gyres – clockwise in the North Atlantic. In the Northern Hemisphere the clockwise loop or gyre is formed with warm water from the Gulf of Mexico (Gulf Stream/North Atlantic Drift) travelling northwards and colder water moving southwards eg the Canaries Current.
- currents from the Poles to the Equator are cold currents whilst those from the Equator to the Poles are warm currents. Cold water moves southwards from Polar latitudes – the Labrador Current. This movement of warm and cold water thus helps to maintain the energy balance.
- ocean currents are greatly influenced by the prevailing winds, with energy being transferred by friction to the ocean currents and then affected by the Coriolis effect, and the configuration of land masses which deflect the ocean currents. Due to differential heating, density differences occur in water masses, resulting in chilled polar water sinking, spreading towards the Equator and displacing upwards the less dense warmer water.

(b) Candidates should be able to name and explain the mechanism of each of the three cells – Hadley, Ferrel and Polar – and should describe their role in the redistribution of energy.

Eg warm air rises at the Equator, travels in the upper atmosphere to c.30°N and S, cools and sinks. Some of this air returns as surface NE or SE trade winds to the Equator to form the Hadley Cell.

The remainder of the air travels north over the surface as Westerlies to converge at about 60°N and S with cold air sinking at the Poles and flowing outwards. This convergence causes the air to rise – some of this air flows in the upper atmosphere to the Poles where it sinks forming the Polar Cell. Candidates may note the Easterlies from the High Pressure area at the Pole.

The remainder of this air in the upper atmosphere travels south and sinks at 30°N and S to form the Ferrel Cell. Credit should be awarded to candidates who recognise that the eastward passage of depressions and associated jet streams deforms any Ferrel Cell out of recognition.

It is in this way that warm air from the Equator is distributed to higher (and cooler) latitudes and cold air from the Poles distributed to lower (and warmer) latitudes.

Question 6 – Lithosphere

(a) Evidence which suggests that Area A on Reference Map Q6 is a Carboniferous Limestone landscape could include:

- extensive areas of bare rock – scars, crags, limestone pavements (eg around and to the west of Malham).
- numerous mentions of underground features such as caves and pot holes (eg those in squares 8570 and 8569) formed due to limestone's susceptibility to solution by acidic rainwater.
- absence of surface drainage over large areas suggesting permeable rock.
- disappearing streams like the one from Malham Tarn which sinks at 894657 and appears to resurface further south below the high cliff of Malham Cove.
- gorges such as Gordale Scar (9164) which, it has been suggested, may have been formed by the massive collapse of cavern roof systems.

(b) Probably the most obvious (sensible!) Carboniferous Limestone feature to choose would be a limestone pavement although some candidates may focus on limestone caves and their associated underground landforms such as stalactites, stalagmites and rock pillars.

In **explaining** the formation of a **limestone pavement**, for example, candidates could refer to such points as:

- the part played by glacial erosion (abrasion) in scraping away any overlying soil cover and thus exposing the horizontally-bedded, rectangular blocks of limestone.
- joints formed in the limestone as it dried out and pressure was released.
- these joints/lines of weakness are more prone to chemical weathering than the surrounding limestone. The limestone is dissolved over time by rainwater (weak carbonic acid) leaving deep gaps (grykes) and intervening blocks (clints).
- continued weathering (both physical and chemical) will further deepen and widen the grykes.

SECTION C

Question 7 – Population

(a) Description could include:
- decreasing birth rate – narrow base and low proportion in youngest age groups.
- decreasing death rate – pyramid narrows naturally from 60's age group and has an inverted pyramid shape lower down.
- increasing life expectancy – many more in the elderly section of the population and a significant number over 100, especially females.
- economically active population decreasing – fewer in the 15 to 65 age groups by 2050.

Explanation could include:
- decreasing birth rate – people wanting smaller (cheaper) families, women following careers, easier family planning/contraception/abortion.
- Decreasing death rate and increasing life expectancy – improved health care, sanitation, housing, food supply, pensions, care for the elderly.

(b) Possible consequences could include:
- the decline in the birth rate may lead to less demand for services/industries needed for the smaller child population with the ensuing problems caused when these are closed or scaled down (eg schools or nurseries).
- the 'greying' of the population may lead to a need for more geriatric care with increased strain/costs on health centres/local authorities/central government.
- a decrease in the economically active population may lead to key jobs not being filled and higher taxes from the working population to pay for their elderly dependents. The qualifying age for pensions may have to be raised and the government may have to reduce their state pension whilst encouraging more private pension plans/health care etc.
- the government may need to encourage the inward migration of key workers that may lead to cultural/language/religious difficulties.
- the government may also need to encourage a higher birth rate.

Question 8 – Urban Geography

(i) For Glasgow candidates may refer to:
- loss of custom for shops in the CBD due to competition from out-of-town shopping centres like Braehead with their large car parking areas.
- consequent closure of shops, especially at the less profitable edges of the traditional CBD due to reduced pedestrian flow, eg High Street end of Argyle Street, giving empty shop units and 'run-down' appearance.
- revitalisation of shopping centres in central CBD – eg building of Buchanan Galleries and renovation of St Enoch Centre in order to compete/keep up.
- shops in CBD may be less overcrowded at peak times, eg Christmas, giving improved shopping experience at these times.
- focus on designer label/high-end shopping taking advantage of CBD status, eg Princes Square, Italian Centre.

(ii) For Glasgow, candidates may refer to:
- pedestrianisation and landscaping of CBD roads eg Buchanan Street, Argyle Street etc to reduce traffic flow in and around the CBD – to increase pedestrian safety and improve air quality and environment. Upgrading of CBD open space, eg George Square.
- diversification of city employment – much greater emphasis on tourist industry (significance of city-break holidays) leading to increased bed accommodation in new CBD hotels (Hilton, Radisson). Hotels can also tap into lucrative conference market given Glasgow's improved image as a tourist and cultural centre.
- alteration of CBD road network – one-way streets (around George Square), bus lanes to discourage use of private transport and encourage use of public transport. Also achieved by increased metering and increased parking charges in and around CBD.
- renovation and redevelopment of many CBD sites to provide modern hi-tech office space (Lloyd's TSB, Direct Line etc) and residential apartments (Fusion Development, Robertson Street).
- building of M8, M77 and M74 extension all designed to keep through traffic off CBD roads.
- younger, more affluent population continues to be attracted to central city area by long-standing concentration of up-market pubs, clubs, cinemas etc (Cineworld in Renfrew Street).
- contraction of number of public transport termini within CBD (2 major railway stations instead of 4) but upgrading of remaining termini, (Buchanan Street bus station, Central Station).

GEOGRAPHY HIGHER ENVIRONMENTAL INTERACTIONS 2009

Question 1 – Rural Land Resources

(a) For a corrie points could include:
- snow accumulates in a (north-facing) hollow on mountainside
- successive layers of snow compress first snowfalls into ice/neve
- ice moves downhill under gravity
- freeze-thaw weathering loosens rock above glacier
- plucking steepens back wall of the corrie
- maximum erosion takes place where weight of ice is greatest
- boulders and stones embedded in ice grind away at the bottom of corrie
- abrasion carving out hollow/armchair shaped depression (over deepening)
- rate of erosion decreases at edge of corrie leaving a rock lip.

Minimum of 3 features needed for full credit.
For an answer to achieve full marks well annotated diagrams must be used.

(b) Responses will vary according to the area chosen but opportunities may include:

tourism, recreation, nature conservation, hill farming, forestry, HEP generation potential, water supply, quarrying. For forestry, candidates could describe the use of plantations for providing wood for furniture, building materials, Christmas trees (economic) and also forest walks, nature trails, picnic sites, mountain bike and orienteering courses (social) with resulting employment (reducing rural depopulation) and profits/taxes for businesses/government.

(c) (i) Conflicts may include:
- traffic congestion especially on narrow rural roads and in car parks - particularly during peak holiday periods
- large volumes of visitor traffic increase air and noise pollution and can spoil the attraction of many villages
- increase in holiday home ownership has pushed up prices forcing many local people to move away
- erosion of footpaths and resulting visual pollution due to high visitor numbers
- some visitors may cause problems for farmers and landowners (eg damage to property such as stone dykes, animal disturbance)
- development of unsightly visitor/leisure complexes/caravan sites etc.

(ii) Solutions might include a variety of environmental conflicts depending on the area chosen but for tourism/traffic it could be:
- traffic restrictions in more favoured areas/at specific peak times, eg one-way streets, bypasses or complete closures
- encourage the use of public transport eg park and ride, minibuses and the use of alternative transport eg cycle paths and bridle ways
- separating tourist and local traffic, the use of permits (for access or parking) in some areas.

Mention of the success/failure of the solution is required for full credit.

Question 2 – Rural Land Degradation

(a) Answers should be able to pick out such points as:
- the considerable fluctuations in rainfall over the period shown
- the preponderance of wetter than average years – certainly between the end of the 19th century and 1970
- the marked concentration of wet years between 1950 and 1970 (only 2 drier than average in these two decades – 1960 and 1969)
- the very obvious pattern of drier years throughout the 1970's and 1980's
- the marked differences in more recent years.

(b) (i) For **Africa north of the Equator** human activities/inappropriate farming techniques contributing to land degradation could include:
- overgrazing; over cropping; deforestation; monoculture; reduced fallow periods; the growing of cash crops; increased population pressure leading to more and more marginal land (more vulnerable to erosion) being cultivated.

For the **Amazon Basin**:
- deforestation – eg for logging/ranching/mineral extraction/road building HEP projects/ resettlement schemes/charcoal burning
- the impact of ranching – forest cleared, used for a few years until grass fails – move and clear a new area of forest and so perpetuate the whole process
- population pressure and shortage of available land arising from these increased demands on the rainforest have resulted in some shifting cultivators returning to tribal lands prematurely and therefore encouraged soil erosion.

(ii) For **Africa north of the Equator** *human* consequences of land degradation could include:
- crop failures/death of livestock – reduced food supply – malnutrition – famine – increased infant mortality rates and death rates
- encourages the drift from rural areas to already overcrowded urban areas – growth of shanty towns
- the collapse of traditional nomadic ways of life

Whilst *environmental* consequences might be:
- soil structure breaks up due to over cropping and monoculture
- wind erosion can remove dried out soil
- deforestation means that soil is exposed and more prone to erosion
- torrential rain often leads to widespread gullying which is impossible to rectify – further loss of precious farmland
- level of water table reduced
- intensified drought due to albedo effect.

For the **Amazon Basin** *human* consequences could include:
- destruction of the way of life of the indigenous population
- clashes between tribal groups and 'outsiders'/developers
- creation of reservations for indigenous people
- impact of "western diseases" on tribal people
- encourages rural-urban migration

Whilst *environmental* consequences might be:
- adverse effect on the rainforests closed nutrient cycle
- leaching of minerals/removal of topsoil/increase in laterisation with loss of protective natural vegetation cover
- increased run-off and flooding; silting up of rivers
- loss of wildlife habitats/biodiversity – some species threatened with extinction
- loss of potentially useful medicinal drugs
- impact on global climate (Greenhouse Effect).

(c) (i) Soil conservation measures employed in **North America** could include:
- contour ploughing
- crop rotation/diversification
- trash farming/stubble mulching
- planting of shelter belts/windbreaks

- strip cultivation
- increased use of irrigation.

(ii) eg in relation to **shelter belts** – trees planted at right angles to the direction of the prevailing wind have managed to reduce wind speeds and provide an effective barrier to protect the soil. The taller and more complete the tree cover, the more effective the shelter. They also improve water retention and help bind the soil together. The negative consequences of this might also be noted eg the area of land occupied by trees/hedges and competition for water and nutrients.

For full credit candidates should mention a minimum of 4 strategies. Some comment on the effectiveness of each measure is also required for full credit.

Question 3 – River Basin Management

(a) Candidates may mention a range of reasons to explain the need for water management including:
- low annual precipitation
- flood control
- regulating flow and storage of water
- power supply for expanding cities and industry
- water for industry
- water for irrigation as food demands increase
- drinking water for increasing population.

(b) Physical factors might include:
- solid foundations for dams
- consideration of earthquake and underground movements
- narrow cross-section to reduce dam length
- large, deep valley to flood behind dam
- impermeable rock beneath reservoir
- sufficient water supply from catchment area
- low evaporation rates, due to small surface area of reservoir
- impact on hydrological cycle.

Human factors might include:
- cost of construction
- proximity of urban areas for water and electricity
- proximity of agricultural areas for irrigation
- cost of displacing people
- cost of compensating farmers and home owners
- impact on communications.

(c) (i) Problems might include:
- difficulties between states which are represented by different political parties
- sharing allocation of water rights
- changing needs of different states including increasing populations and increasing irrigation
- increased pollution and salinity downstream affecting water quality
- shared costs of purification and desalination plants
- impact of dam construction on consumers downstream
- relationship between neighbouring countries.

(ii) These problems could be overcome by having political agreements eg the Colorado River Compact which divided up water allocations based on historical rainfall patterns. International agreements may be needed where different countries are involved.

(d) Social benefits include:
- improved water supply for drinking
- irrigation providing an increased food supply
- less water borne disease
- population increases sustainable
- greater availability of electricity
- opportunities for tourism and recreation
- increased fresh water improves health and sanitation.

Economic benefits include:
- improved navigation and roads across dams
- HEP and water for industry
- irrigation allowing double cropping and commercial farming
- income from tourism and recreation.

Environmental benefits include:
- flood control
- reliable seasonal water supply
- dramatic scenery around dams and reservoirs
- introduction of new wildlife habitats.

Candidates must refer to all 3 sections for full marks

Question 4 – Urban Change and Management

(a) Answers will depend upon the EMDC chosen, but for Spain answers might suggest:
- concentrations on and around the coast of Spain, both on the North coast and along the East and South coastal areas
- coastal cities would include ferry terminals (Santander, Palma), historical trading ports (Barcelona) and holiday areas accessed by airports (Malaga, Alicante)
- along rivers for communication, trade, raw materials (Sevilla, Zaragoza)
- major cities on Spanish islands (Palma, Mallorca and Las Palmas, Gran Canaria)

(b) (i) Social, Economic and Environmental problems should be related to the candidate's chosen city. Answers would be enhanced by convincing relevant details on the chosen city such as **named** shanty areas or specific projects to tackle problems.

Problems might include:
- continued growth of these shanty towns (favelas, bustees etc) in and around the city
- shanty areas are characterised by poor quality home-made dwellings, overcrowding, inadequate water and power supplies, poor sanitation, disease and general lack of amenities like services, schools and hospitals
- they are often sited on unstable hillsides, marshy areas or other areas avoided by other building
- unemployment or underemployment and poor wages for the few jobs available
- 'grey' or 'black market' economies with problems of drugs and high crime rate
- chronic traffic congestion and high pollution levels from nearby industries
- sites are illegally settled and may be bulldozed and removed by city authorities at any time.

(ii) Ways to tackle problems might include:
- self-help schemes (eg São Paulo) where city authorities provide basic housing made of breeze block and roof tiles. Local residents supply the labour for 'finishing off' and digging ditches for water supply and sanitation
- basic amenities such as power, clean water, roads and community facilities may be provided
- groups of residents within shanty town areas may form community groups to share trade skills to improve existing facilities within the larger shanty town
- city authorities may build high-rise apartment blocks in suburbs to provide high-density housing to replace the extremely high-density living in shanty areas

Some qualitative statements on the success or otherwise of these schemes is required to attain full marks. For example, "the advantages of self-help schemes are that costs are kept to a minimum to maximise the number of 'basic shell' houses that can be built. Working together can establish community spirit with shared common purpose."

(c) (i) Maximum of 14 in (b) (i) and (ii) – if named city.

Traffic congestion in an EMDC city. For Aberdeen candidates might suggest:

- increased commuting from dormitory towns and villages to N, W and S of Aberdeen as people seek quieter living conditions focuses rush hour traffic on major traffic junctions eg Haudagain roundabout
- Aberdeen has major industrial areas at Altens and Dyce, leading to large commuter flows outwith the city centre
- major roads have to converge to cross the rivers Don and Dee, leading to bottlenecks at bridges over the rivers
- around 15,000 journeys per day in Aberdeen are generated by through traffic, clogging up city streets unnecessarily
- more stringent traffic regulations in and around the CBD and shortage of car parking facilities, leading to unnecessary traffic flow as spaces are sought
- growing car ownership, related to high disposable incomes and increased number of 2 (or more) car families. > 60% of employed people in Aberdeen travel to work by car
- increased use of private transport to do 'school run' during rush hours, due to safety issues
- increase in number and size of lorries and buses which often find it difficult to manoeuvre in outdated road network, delaying other traffic
- increase in need for road maintenance due to increased traffic flow and weight of modern lorries
- shutting off of side roads formerly used as 'rat runs' focuses all traffic on to main arterial roads.

(ii) Candidates may suggest protests and land-use conflicts that would be due to:

- breaching of green belt land
- roads such as the AWPR use up large tracts of land, often good quality farmland or recreational land, leading to protests by groups such as the Aberdeen Greenbelt Alliance or Friends of the Earth who want to preserve green belt areas
- removal of sensitive woodlands and meadows in the Dee Valley may harm endangered species in the areas, such as the Red Squirrel
- once roads such as the AWPR have been built, they act as a focal point for developers wanting to build houses and/or industrial areas and/or out-of-town retail parks with vast buildings and huge car parks taking up large areas of land, these all benefit from the improved access brought by the new road, these also in turn lead to further breaching of the green belt land, including potential loss of golf courses, country parks etc
- property blighting, where any properties which are within sight or sound of the road may lose much of their value
- spiralling costs of projects like the AWPR lead protesters to argue that the huge amounts of money involved could be used to improve existing infrastructure and modernise public transport
- compulsory purchase orders are placed on properties on the proposed route and people are forced out of their houses.
- farms are often split by such roads, causing access problems for farmers and their livestock
- encroachment of housing may increase crime and vandalism for farmers and their property.

Question 5 – European Regional Inequalities

(a) Candidates should note the higher levels of development in pre-2000 states. This may be due to locations near the economic Core, the length of membership and subsequent financial benefits, the quality of infrastructure, the proximity of markets, the ease of trade and the sourcing of resources. The relative prosperity of the pre-2000 states also leads to a better quality of life. Candidates will also gain credit from noting anomalies and where possible outlining reasons for these anomalies eg Portugal and Slovenia.

(b) (i) Credit should be awarded for candidates noting that Objective 1 status is awarded to Europe's peripheral areas. A maximum of 6 marks should be awarded for identifying areas or countries eg Ireland, SW England, Wales, Portugal and Spain, S Italy, Greece, Sweden, Finland, the eastern part of Germany and the new states of the Eastern bloc.

(ii) Countries would benefit from Objective 1 status through the following:

- support for infrastructure improvements
- support for employment training and education
- support for production/manufacturing sectors
- environmental protection
- improving access to the peripheral areas
- improving IT, literacy and numeracy.

(c) (i) Answers will be dependent upon the country chosen but must be authentic for the candidate to score full marks.

Physical factors could be related to (for example):

- difficult terrain/relief (mountain ranges: Northern Spain, Central Italy, Highlands of Scotland)
- physical isolation (Highlands of Scotland/Wales, areas in Central Spain, much of Southern Italy)
- climatic problems such as drought (Southern Italy, Central, Southern and Eastern Spain, Greece)
- prolonged winters (Northern Sweden, much of Finland)
- natural disasters (earthquakes; Greece, Italy; volcanic eruptions; Southern Italy).

Human factors could include (for example):

- differences in employment opportunities
- decline in the range and scope of opportunities in the rural-based economy
- decline in range of relevant skills in a declining industrial area
- perceptions of inward investors
- problems of land ownership and tenure
- political/terrorist factors (ETA)
- overdependence on seasonal employment in (for example) the tourist industry
- variations in investment infrastructure
- distance from main markets/Euro-core.

(ii) Once again answers will depend upon the country chosen but national government strategies to tackle regional inequalities might include:

- regional development status, Enterprise Zone status, capital allowances, training grants, assistance with labour costs
- specific assistance to former coal mining/iron and steel areas
- intervention of national government resulting in the location of major government employers in disadvantaged areas DVLA (Swansea), MOD (Glasgow), SNH (Inverness)
- intervention by national government to encourage inward investment – particularly in newer industries – electronics/call centres.

Question 6 – Development and Health

(a) Essentially single indicators are too broad/generalised:
- they are averages which disguise or distort wide internal variations eg a few immensely wealthy families but the majority of the population may be living at subsistence level
- combining indication on health, education and the economy give a more balanced view of development
- some regions/areas of a country may be much better off than others – 'north-south' or 'urban-rural' contrasts
- GNP figures are in some cases inflated by oil revenues (showing a big gap between these and other indicators that have yet to 'catch up')
- subsistence agriculture and 'barter economies' are not included in wealth indicators
- certain indicators are perhaps irrelevant to the real quality of life in many poorer ELDCs eg TVs per household when there is no electricity supply.

(b) Differences in the levels of development between ELDCs (Economically Less Developed Countries) may be due to:
- mineral reserves eg Saudi Arabia and similarly positioned Middle East countries with vast reserves of oil. They also have stable (if despotic) government regimes/monarchies that leads to the generation of huge wealth. This wealth can 'trickle down' to a wide sector of the population. Other countries may have no reserves of minerals in demand by the EMDCs eg Burkina Faso
- political instability eg many have unstable regimes or are suffering from border wars and/or civil wars eg Zimbabwe, Sudan and Indonesia
- colonial links eg some Caribbean countries receive support from European countries because of their former colonial ties
- strategic locations eg South Korea and many Central American countries receive additional support from leasing land for military bases
- encouragement of entrepreneurial skills and the ability to attract in major world companies eg by offering an educated, resourceful and relatively cheap work force (South Korea) and/or incentives eg 10 years rent free factory sites in Vietnam
- natural disasters eg Bangladesh (cyclones and floods), Niger (recurring drought and associated famines) will limit progress as development money is spent on repairing infrastructure and humanitarian aid.

(c) (i) For **Malaria** – Environmental factors:
- suitable breeding habitat for the female Anopheles mosquito – areas of stagnant water to lay eggs in
- hot and wet climates such as those experienced in the Tropical Rainforests or Monsoon areas of the world
- temperatures between 15°C and 40°C
- areas of shade in which the mosquito can digest blood.
Human factors:
- nearby settlements to provide a 'blood reservoir'
- encouraged by bad sanitation and poor irrigation or drainage that leaves standing water uncovered eg tank wells, irrigation channels, water barrels, padi fields.

(ii) Strategies used to combat the spread of Malaria can include:

Trying to **eradicate** the mosquito:
- insecticides eg DDT and now Malathion
- mustard seeds thrown on the water that become wet and sticky so dragging the mosquito larvae under, drowning them
- egg-white sprayed on the water creates a film which suffocates the larvae by clogging up their breathing tubes

- bti bacteria grown in coconuts – the fermented coconuts are broken open after a few days and thrown into the mosquito infested ponds. The larvae eat the bacteria and have their stomach linings destroyed
- larvae eating fish introduced to ponds
- draining swamps, planting eucalyptus trees that soak up excess moisture, covering standing water
- genetic engineering eg of sterile males

Treating those suffering from malaria:
- drugs like chloroquin, larium and malarone
- quinghaosu extracted from the artemesian plant – a traditional Chinese cure
- continued search for a vaccine – not available as yet
- education programmes in –
  - the use of insect repellents eg Autan
  - covering the skin at dusk when the mosquitoes are most active
  - sleeping under an insecticide treated mosquito net
  - mesh coverings over windows/door openings
- WHO 'Roll back malaria' campaign
- the Bill and Melinda Gates Foundation

(iii) Benefits to ELDCs of controlling disease may include:
- funds can be diverted elsewhere in the Health sector or transferred to other budgets that help development
- national debt can be reduced
- the workforce will be fitter (eg farmers better able to produce food), thus also helping to raise health levels
- productivity will increase as the workforce takes less sick leave/life expectancy increases
- the area will become more attractive to tourists, foreign currency income can be generated and this will also assist in developing tourism related services/industries
- a possible reduction in birth rates as a result of a fall in infant mortality rates.

(d) Examples of Primary Health Care (PHC) strategies may include:
- use of barefoot doctors – trusted local people who can carry out treatment for more common illnesses – sometimes using cheaper traditional remedies
- use of ORT (Oral Rehydration Therapy) to tackle dehydration – especially amongst babies. This is an easy, cheap and effective remedy for diarrhoea/dehydration
- provision of vaccination programmes against diseases such as polio, measles, cholera. Candidates may also refer to PHC as based on generally preventative medicine rather than (more expensive) curative medicine
- the development of health education schemes in schools, community plays/songs concerning AIDS, with groups of expectant mothers or women in relation to diet and hygiene. Oral education being much more effective in illiterate societies
- sometimes these initiatives are backed by the building of small local health centres staffed by doctors (like GPs)
- PHC can also involve the building of small scale clean water supplies and Blair toilets/pit latrines – often with community participation
- The use of local labour and building materials is often cheaper, it also provides training/transferable skills for the participants and gains faster acceptance/usage in the local and wider community.

Explanation of why PHC is more effective in ELDCs is required to gain full marks.

GEOGRAPHY HIGHER PHYSICAL AND HUMAN ENVIRONMENTS 2010

SECTION A

Question 1 – Atmosphere

(a) Human factors
  - Carbon Dioxide: from burning fossil fuels – road transport, power stations, heating systems, cement production and from deforestation (particularly in the rainforests) and peat bog reclamation/development (particularly in Ireland and Scotland for wind farms).
  - CFC/PFCs: from aerosols, air-conditioning systems, refrigerators, polystyrene packaging etc.
  - Methane: from rice paddies, animal dung and belching cows.
  - Nitrous oxides: from vehicle exhausts and power stations.
  - Sulphate aerosol particles and aircraft contrails: global 'dimming' – increase in cloud formation increases reflection/absorption in the atmosphere and therefore cooling.

  NB There were 6 man-made greenhouse gases included in the Kyoto protocol (Carbon Dioxide, Methane, Nitrous Oxide, Hydroflurocarbons, PFCs and Sulphur Hexafluoride). Many more powerful than CO_2.

(b) Answers may include:
  Melting of the ice sheets/glaciers
  - A rise in sea level with subsequent migration as islands and coastal areas are submerged. Loss of plant and animal habitats in these areas eg impact on polar bears which could lead to a loss of tourism/more problems in settlements as the bears scavenge instead of hunting on the ice.
  - New transportation routes across the Arctic Ocean ie the North West Passage with resulting benefits to trade/previously ice bound coastal settlements.
  - Extension of mineral exploitation into the Arctic with positive and negative consequences.

  Changing rainfall/temperature patterns
  - Higher or lower rainfall/temperature and maybe more extreme weather depending on where you are with resulting increasing/decreasing crop yields, more floods/drought/hurricanes/tornadoes etc.
  - Extension or retreat of vegetation (and associated wildlife) by altitude as well as latitude – growing vines/sunflowers in Scotland, spread of malaria, the loss of the Cairngorm Arctic habitat etc.
  - Change in ocean currents (El Nino/La Nina).
  - Change to the Atlantic Conveyor – disruption of the thermohaline circulation.

Question 2 – Lithosphere

A sequence of diagrams, fully annotated, could score full marks in either part (a) or part (b).

(a) Conditions and processes which encourage the formation of scree slopes.
  - Steep and bare rock faces with lines of weakness/well-joined carboniferous limestone.
  - Cold climate where temperatures often fall below freezing point at night.
  - The two factors above allow physical weathering to take place in the form of freeze-thaw action/frost shattering, where water collects in the rock fractures, freezes and expands by about 9% exerting great pressure on even the hardest rock.

  - Repeated freeze-thaw action splits the rock into large sharp fragments which break off and are moved downhill by gravity to accumulate at the base of steep slopes as a scree or talus slope as large heaps of rock debris.

Full marks can only be achieved if one or more annotated diagram is included.

(b) Processes involved in formation of a corrie.
  Corries
  - Snow accumulates in mountain hollows when more snow falls in winter than melts in the summer.
  - North/North-east facing slopes are more shaded so snow lies longer.
  - Accumulated snow compresses into neve and eventually ice.
  - Plucking, when ice freezes on to bedrock, fractures it and incorporates it into the glacial ice.
  - Abrasion, when the angular rock within the glacial ice grinds away the valley sides and floor, over-deepening the hollow along with rotational movement of glacier.
  - Glacier moves downhill due to gravity.
  - Rotational movement not so powerful at corrie edge, allowing rock lip to form which traps water as ice melts, leaving a lochan.

Question 3 – Population Geography

(a) Answers could include points such as:

Description	Explanation
The largest number of migrants come from Poland (124,000)	Due to the expansion of the EU in 2005 and freedom of movement for workers
The second highest source was India (about 100,000)	Possibly due to Commonwealth links or people who already have family in the UK's large Indian sector
A large proportion (48,000) came from Australia	Due to the lack of a language barrier and the increasing trend towards young people travelling for work experience

Reference can also be made to relevant push and pull factors.

(b) Answer will depend on migration studied.
  A developed answer will refer to both advantages and disadvantages for the country.
  From Mexico into California, answers may include:

  Mexico

Advantages	Disadvantages
• The pressure on resources and jobs was lessened. • The birth rate was also lowered as most migrants were of childbearing age. • Money was often sent back to the families left behind, which helped to stimulate the economy – it is Mexico's biggest source of foreign income ($6 bn per year). • When migrants return, they can bring back new skills, which can be used in the donor country.	• The active population left, creating a burden on the economy. • Those most educated left creating a 'brain-drain'. • Families are divided as males leave. • Death rate increases as an elderly population is left. • In the long term this creates dependency upon money sent back to home villages.

California

Advantages	Disadvantages
• The short-term labour gap was filled – migrants filled jobs Americans did not want. • Mexican culture has enriched the border states with language, food and music. • Increased population leading to increased taxation levels. • Labour costs reduced – agricultural sector benefits from this.	• Migrant workers feel discriminated against and socio-economic problems have ensued. • When recession hit in the 1980s unemployment rose and racial tension was exacerbated. • Ghettos developed in the poorest districts. • TB has increased along the border. • Illegal migration costs the USA millions of dollars for border patrols and holding centres.

Question 4 – Urban Geography

(a) The following characteristics may be noted:
- Densely packed and irregular street pattern.
- Transport centres eg bus station and railway station.
- Bridging points across River Ouse.
- Historical buildings eg The Minster, Castle.
- Important buildings eg information centre, churches and Town Hall.
- Evidence of inner ring road.

(b) The advantages of the residential location and environment may include:
Area B (suburban housing area – Rawcliffe).
- Access to A19 for commuting to CBD.
- Near park and ride for commuting, and National cycle route.
- Modern design of cul-de-sacs and crescents for privacy and preventing through traffic, and roundabouts at access points.
- Services including a church for local use.
- Near industrial estate GR593553 for employment.
- Tourist facility to east ie Nature Reserve, and caravan site.
- Attractive environment ie small lake, on edge of town near farmland.
- Clifton Moor Retail Park 591558.

Area C (commuter village – Copmanthorpe).
- 6km from centre of York for shopping, work and entertainment.
- Nearby sliproad onto A64, ideal for commuters.
- Small, quiet village with a few services eg post office, public house, church.
- Leisure facility to north ie golf course.
- Environmentally attractive with Ebor Way going through the village and Askham Bogs Nature Reserve to the north.
- Surrounded on three sides by farmland.

(c) Candidates should be able to demonstrate an understanding of the issues which arise when 'urban' land uses invade a previously 'rural' area.
Land uses which would conflict with further expansion would include
- National walking and cycle trail.
- New shopping centre – expansion may be restricted.
- Leisure facilities eg racecourse, golf course.
- Various farms eg White House farm.
- A64 bypass.

- Accommodation including Manor Hotel, caravan and camping site GR600476.
Other land uses include forestry, small villages, college, university, electricity transmission lines.

SECTION B
Question 5 – Hydrosphere

Full marks can only be achieved if one or more annotated diagram is included.

(a) The explanation should include 8 points, all of which could be included in a well annotated diagram.
Points could include:
- Development of pools and riffles (differences in speed and depth).
- Erosion on the outside (concave bank) of bends due to faster flow.
- Helicoidal flow removing material.
- Deposition on the inside (convex bank) due to slower flow.
- Formation of point bars.
- Migration of meanders downstream.

(b) The physical factors may have included:
- The close proximity of a tidal limit to York, GR594445.
- A low lying floodplain around or below 10 metres in height.
- Tributaries joining the River Ouse eg River Foss (605510).

The human factors may have included:
- Home building on a flood plain.
- River straightening.
- Man-made strengthening and narrowing of river banks.
- Facilities deliberately located on floodplain eg racecourse, caravan and camping site.
- Land use changes in catchment.
- Increased run-off from the 'urbanised' environment.

Question 6 – Biosphere

Students may answer this question using the headings on the graph or the more usual progression from strandline to climax vegetation.

- **Plant cover increases** – the amount of sand showing through the dune decreases as more of the ground is covered by vegetation. Little cover in the pioneer stage, more in the building stage and complete cover in the climax stage unless disturbed by animals/humans/storms etc.
- **Soil moisture increases** – rain/fresh water is trapped with added humus/plant cover and longer rooted plants drawing water up from the water table. Xerophytic plants found in the drier strandline (sea sandwort, sea rocket, saltwort) and embryo dune (sea or sand couch, lyme grass, frosted orache). On the yellow dune Marram grass has long rhizomes to spread through the sand. Dune slacks at or near the water table have hydrophytic species like reeds, rushes and flag iris.
- **Organic matter content increases** – decaying pioneer species adding humus to the sand. In the fore-dune more plants stabilise the sand adding humus (sea bindweed, sea holly, sand sedge, and marram grass) changing the sand to a sandy loam and from the sandy colour of this and the yellow dune to the grey dune.
- **PH decreases** – shells ($CaCO_3$) producing alkaline conditions on shore, more neutral pH by the climax stage as plants decay and add acid to the soil. The grey dune plants include sand sedge, sand fescue, bird's foot trefoil, heather, sea buckthorn and grey lichens. In the climax stage a range of plants from heathers to birch, pine or oak woodland can grow depending on the final pH value of the soil.
- **Salinity decreases** – Increased distance from the sea and salt water/tides/spray increases the amount and variety of plant species that can cope with the conditions.

SECTION B

Question 7 – Rural Geography

(a) Candidates may answer this question in one or two parts. Answers could include the following:

(i) Mechanisation increases the efficiency on a farm enabling the farmer to plough, sow, spray, etc more quickly, covering larger areas. It also speeds up harvesting and results in the product being delivered to markets fresher and at a higher premium (eg Bird's Eye peas). It also allows for a smaller work force and therefore lower wage bills for the agribusiness usually involved in these farms. It allows for the use of satellite technology/ computers to control the application of fertilisers to particular areas of fields to improve yields (yet decrease the cost and waste) as only the required amounts are delivered to each segment according to the soil quality there.

(ii) **People in the environment** – this leads to depopulation and derelict buildings, deserted rural villages (eg on the Great Plains).
Farm sizes – increasing 'agribusiness' type farming with amalgamated farms, larger fields, fewer hedgerows or boundaries to allow for machinery and increase yields.
More or larger buildings for storage of machinery.
Pollution – air and water pollution from machinery itself (exhaust fumes/noise/accidents with diesel) and run-off from over application of fertilisers.

(b) For EU Policies eg Single Farm Payment:
This payment has replaced the existing support schemes to farmers like the arable area payment scheme and includes other entitlements like the set-aside entitlement.
The total payable to the farmer is calculated using the historical payments made to the farm from 2000 to 2002. The farmer gets a single payment based on these calculations. To continue to receive this payment the farmer must keep the land in 'good agricultural and environmental condition' – this includes an obligation to keep land in set-aside. Penalties will be made if these conditions are not met – government inspectors will visit to check!

For Genetically Modified food/genetic engineering:
Scientists manipulate the genes of plant cells by adding/deleting DNA. The first commercial genetically engineered food was the Flavr Savr tomato – by adding a fish gene it had a longer 'shelf-life'. Since then the developments have concentrated on four main crops – oilseed rape, cotton, maize and soya bean. These crops have been made herbicide resistant (they can tolerate the herbicides that will kill the weeds) and/or pest resistant (they produce a toxin that kills the pest that would normally eat them). This technology has also been used to improve the taste, nutrition or drought-resistance of the crop. Ethical concerns, health risks, environmental concerns eg cross-pollination to produce 'superweeds' have led to these crops being limited in area (although that is still estimated at 1 million square kilometres).

Question 8 – Industrial Geography

(i) Answers will depend on the chosen area. For the industry in the question.
- Increased competition from overseas markets, particularly NICs.
- Increased competition leading to falling prices and profits.
- Falling customer demand for the product as new technology takes over.
- Cheaper labour from countries such as India.
- Improved (and cheaper) transport and communication means that products do not need to be manufactured near to the market.
- Ending of government incentives to encourage new industries.
- Modernisation of plants in order to compete can result in redundancies.
- Rationalisation of company leading to overseas plants being closed.

(ii) Again, answers will depend on the area chosen but will include effects such as:
- Associated service and supply industries close.
- Cycle of economic decline sets in.
- Depopulation, particularly amongst young people and young families.
- Leading to further service closures.
- Areas in decline find it difficult to attract new investment as area becomes run down.
- Rise in cases of depression.
- Rise in crime rates in area.

GEOGRAPHY HIGHER ENVIRONMENTAL INTERACTIONS 2010

Question 1 – Rural Land Resources

Well annotated diagrams will be awarded full credit.

(a) Candidates should refer to the process of coastal erosion and deposition within their answer, ie hydraulic action, abrasion, solution, attrition and wave movement up/down beaches with longshore drift.

A typical answer for a cave/arch/stack may include:
Caves are most likely to occur where the coastline consists of hard rock and is attacked by prolonged wave attack along a line of weakness such as a joint or fault in the rock. The waves will attack the line of weakness by abrasion, hydraulic action or solution. Over time, horizontal erosion of the cave may cut through the headland to the other side, and form an arch. Very occasionally a blowhole will be created within the cave where compressed air is pushed upwards by the power of the waves and vertical erosion occurs. Continued erosion of the foot of the arch may eventually cause the roof to collapse leaving a stack, isolated from the cliff. This in turn will be eroded yet further to leave a stump.

(b) Candidates must discuss at least three land uses to achieve full marks.
Ideally candidates should identify the specific feature of the landscape and then go on to explain the opportunity it provides. Responses will vary according to the area chosen but opportunities might include:
Social – tourism, recreation, nature conservation.
Environmental – farming, forestry, energy generation (wind/waves/tidal), quarrying.
Industry – ports, oil industry.

(c) Candidates should be able to discuss both sides of the argument in this development, to achieve full marks.

Advantages/People arguing for the plan	Disadvantages/People arguing against the plan
Jobs will be created during and after the construction.	Labour force – probably not local people, strain on local services.
The golf course will boost visitor numbers and bring money into the economy of the local area.	This will damage the mobile sand dunes and associated wildlife. Loss of habitat/biodiversity.
The new houses will attract people to move here and boost local services like schools.	Locals/wildlife enthusiasts will no longer have access to the beach/sand dunes on the estate.
Local businesses will benefit by supplying the hotel ie taxi companies, food/farm contracts.	Car parks/roads and buildings associated with the development will cause visual pollution.

(d) To gain full marks candidates must comment on the effectiveness of their solutions/measures taken to resolve environmental conflicts.

Measures taken to resolve environmental conflicts might include:
• Traffic restrictions in more favoured areas.
• Reducing congestion on busy roads using a one-way system.
• Encouraging the use of minibuses.
• Seperating local and tourist traffic.
• Attempting to develop wider spread of 'honeypot' areas.
• Providing cheap local housing for inhabitants of area.
• Screening new buildings, car parks etc behind deciduous trees and using only local stone for buildings.
• Better visitor education.

Question 2 – Rural Land Degradation

(a) • **Rainsplash** – usually the first stage in the erosion process, the impact of raindrops on the surface of a soil causing the soil particles to be moved. On steeper slopes they move further downhill. This means that resettled sediment blocks soil pores resulting in surface crusting and lower infiltration.
• **Sheet erosion** – the removal of a thin layer of surface soil which has already been disturbed by rainsplash, accounts for large volumes of soil loss, rarely flows for more than a few metres before concentrating into rills. Typically results in the loss of the finest soil particles which usually contain the nutrients and organic matter.
• **Rill erosion** – small eroded channels, only centimetres (up to about 30cm) deep and not permanent features, often obliterated by the next rainstorm, or develop into gullies.
• **Gully erosion** – steep sided water channels, several metres deep which can cut deeply into the soil after storms and are often permanent. Rain water running into the gully scours the sides or undercuts the head wall which results in the gully migrating. Widening of gully sides can occur by undercutting or slumping.

(b) Human causes of land degradation will vary according to the location chosen but may include reasons such as:

North America
For the **Dust Bowl**:
• Use of techniques better suited to the moister eastern states.
• Monoculture, especially of wheat or demanding crops (cotton), depleted the soil of moisture and nutrients.
• Deep ploughing of fragile soils (previously these had been held in place by natural grasslands).
• Marginal land ploughed – particularly in wet years – leaving them in a fragile condition in dry years.
• Ploughing downslope creating opportunities for rill erosion.
• Farm sizes being too small so forcing farmers to overcrop – particularly when prices were low and therefore income was low.

For the **Tennessee Valley**:
• Much of the area was cleared of its trees – this opened up the soil surface to erosion.
• Mining and farming also cleared the natural vegetation and led to soil erosion.
• The farmers cultivated steep slopes which were ploughed up and down the slope.
• Overcropping had already weakened the soil.
• The eroded soil was dumped in rivers and this caused them to flood.
• A lack of fertilizer caused the soil to lose its structure and become vulnerable to erosion.

For **Africa, north of the equator,** mention might be made of overgrazing, overcropping, deforestation, monoculture, farming cash crops.
• Allowing more grazing than the pasture can support (eg in West Africa) where herd size is a status symbol.
• Allowing the soil to be stripped bare leaving it vulnerable to erosion.
• Increased population density caused by falling death rates leading to overcultivation.

- Deforestation for firewood/building.
- Bush fires to clear land for farming.
- In some places peasant farmers have had to farm marginal land due to the best land being used for cash crops (eg in parts of Sudan).
- The drought may have caused nomads to move into villages where the land may now be over-cultivated (eg in Burkina Faso).

For the **Amazon Basin,** answers will be based on deforestation:
- Deforestation – for eg ranching/mineral extraction/ logging/road building/poor peasant farmers.
- Loss of protective cover of trees due to deforestation.
- This allows heavy tropical rainfall to erode the soil.
- Exposure to increased sunlight due to deforestation leads to the soil baking and becoming useless.
- The loss of the root system which previously bound the soil together.
- Deforestation also leads to increased leaching of the soil rendering it useless in addition to erosion.
- The impact of ranching: forest cleared, used for a few years until grass fails – move and clear a new stretch of forest and continue the process.

(c) For **Africa, north of the equator** descriptions may include:
- Crop failures and the resulting malnutrition leading to famine eg Sudan, Ethiopia and much of the Sahel.
- Southward migration on a large scale – usually into shanties on the edge of the major cities.
- The collapse of the nomadic way of life due to the lack of grazing and water.
- Many nomads forced to settle in villages – with a consequent increase in pressure on the surrounding land.
- The breakdown of the settled farmer/nomad relationship in places like Yatenga province in Northern Burkina Faso.
- Disease and illness can become endemic.
- Conflict within countries as people move and re-settle.
- Countries increasingly rely on international aid.

For the **Amazon Basin** answers may include:
- Destruction of the way of life of the indigenous people eg clashes between the Yanomami and incomers.
- Destruction of the formerly sustainable development eg rubber tappers and Brazil Nut collectors.
- Clashes between various competing groups eg the violent death of Chico Mendez allegedly at the behest of ranchers.
- Reduction of fallow period leading to reduced yields with obvious consequences for the dependent population.
- Creation of reservations for indigenous people.
- Increase in 'western' diseases.
- Increase in alcoholism amongst indigenous population.
- People have been displaced and forced into crowded cities ending up living in favelas.

(d) Answers should be able to give reasonably detailed information about farming methods, and must include some explanation of these methods, for example:
Shelter belts – on low lying land affected by strong winds shelter belts are rows of trees grown across the direction of the prevailing wind. They act as a barrier to slow down winds and protect the soil. The taller and more complete the barrier of trees the more effective the shelter.

Other farming methods might include:
- Crop rotation.
- Diversification of farming types.
- Keeping land under grass or fallow.
- Trash farming/stubble mulching.
- Replanting shelter belts.
- Strip cultivation and intercropping.

- Increased irrigation.
- Soil banks by keeping soils under grass rather than ploughing.
- Diversification by farmers into recreation.
- Contour ploughing.
- Terracing.
- Use of natural fertilisers.

Question 3 – River Basin Management

(a) (i) Description and explanation of pattern of river flow might include:
- **Description** – very irregular flow from month to month from 2,000 cumecs in Jan/Feb to 13,000 cumecs in Aug/Sept.
- Similar pattern from year to year with peaks and troughs at the same time each year.
- **Explanation** should refer to the fact that troughs relate to dry months from Dec-Mar while peaks occur after heavy rainfall of Jun/Jul/Aug. Rainfall monthly figures indicate monsoon rains.
- Discharge increasing before heavy rainfall suggests discharge is fuelled by snow melt from surrounding mountains shown on reference map Q3A.

(ii) Description and explanation of need for water management might include:
- Reference map Q3A indicates that the Irrawaddy River has many tributaries and the river basin has a very high drainage density leading to unpredictability of river flow which is dependent on when and how quickly snow melts in surrounding mountain areas.
- Rapidly increasing population in Myanmar gives increasing demand for water for domestic, power, industrial needs.
- Increasing demands from farmers for irrigation water to try and feed increasing population.
- Rainfall graph for Myitsore indicates seasonal nature of rainfall – extremely dry from November to April but huge monthly figures for June/July/August – leading to flooding and also run-off of water that could be stored and used in dry months.
- Temperature graph for Myitsore indicates hot temperatures throughout the year leading to very high evaporation rates. Monthly temperatures peak at 35°C.
- Reference diagram Q3A indicates that there is a need to regulate flow of river to prevent flooding during peak discharge and to keep water level high enough for navigation in dry months.

(b) Physical factors might include:
- Geologically stable area away from earthquake zones/fault lines.
- Solid rock foundations for weight of dam.
- Narrow valley cross-section to reduce dam length.
- Large, deep valley to flood behind dam to maximise amount of water storage.
- Lack of permeability in rock below and around reservoir to prevent seepage.
- Low evaporation rates.
- Large catchment area above dam to provide reliable water supply.

(c) Answers should be authentic for the chosen river basin. Candidates must refer to all 6 sections for full marks.

Answers will depend on the river basin chosen. However, for the Colorado River they might include:

Social benefits:
- Fresh water supply for growing desert cities eg Phoenix.
- Better standard of living in hot, dry climate with air conditioning, swimming pools, landscaping etc.
- Areas at reservoirs, eg Lake Mead, give opportunities for tourism, water sports, fishing etc.
- Regulation of river greatly improves flood control on river.

Social adverse consequences:
- People had to be moved off their land as valley areas were flooded.
- Loss of burial sites and other Native American sacred areas.
- Disagreements between states and countries with regard to allocation of water from river.

Economic benefits:
- Cheap HEP attracted industries eg electronics to take advantage of the area's cheap land and low taxes.
- Benefited tourist industry with reliable water supply – attractions like the Grand Canyon, gambling in Las Vegas, Hoover Dam etc.
- Expansion of irrigated land led to agribusiness-style farming.

Economic adverse consequences:
- Huge cost of building the dams eg Central Arizona Project cost $6 billion.
- High cost of maintaining dams, power plants and irrigation channels.
- Subsidised water for farmers has led to water wastage and the growing of crops that could be produced cheaper elsewhere.

Environmental benefits:
- Reservoirs provide sanctuaries for waterfowl and wading birds like the blue heron.
- The National Recreation Area around Lake Mead has more than 250 species of birds.
- Reliable seasonal water flow for plant and animal life.

Environmental adverse consequences:
- Water in river and on farmland becomes saline with high evaporation rates – farmers downstream have to switch to more salt-tolerant crops.
- Change in river regime has caused the loss of many animal habitats eg the drying up of Colorado delta area where there used to be a great variety of birdlife.
- Huge amounts of water loss by seepage through the sandstone rocks around Lake Powell. Scenic attractions like the Rainbow Bridge are being affected by the high water levels in Lake Powell.

Question 4 – Urban Change and its Management

(a) Candidates should be able to identify the overall increase in number of megacities from 1975 to 2015.
They should recognise that the greatest area of growth is in developing countries rather than developed countries.
Within this they should identify patterns of increased growth in particular areas of the world eg India and China in Asia and USA and Brazil in the Americas.

(b) (i) Candidates should be able to demonstrate authentic knowledge of a city they have studied. Answers could include the following points:

Rural push:
- Low income from farming and related work.
- Lack of employment in manufacturing and service industry.
- Lack of education and low literacy levels.

- Poor health facilities and higher levels of disease, malnutrition etc.
- Low quality of life, poor sanitation, lack of electricity.
- Poor quality of infrastructure.
- Resettlement, civil unrest, environmental degradation.

Urban pull:
- Industrial employment, both manufacturing and service.
- Informal opportunities for employment.
- Increased income.
- Better housing, education, health facilities.
- Improved infrastructure.
- 'Bright lights' ambitions.

(ii) Candidates should relate socio-economic and environmental problems to their chosen city.
- Impoverished and overcrowded areas of the city which lack many public utilities and amenities of water supply, electricity and sewerage.
- Semi-urban peripheral districts with poor housing quality and poor economic opportunities. Squatter settlements are located on steep upland areas. Areas lack basic services eg schools, piped water and hospitals.
- High incidence of disease.
- High rates of unemployment and growth of 'grey' economy and black market.
- Crime, drugs and prostitution.
- Problems of waste disposal include open sewers, toxic industrial waste contaminating water supply, lack of refuse collection and landfill sites for solid waste.
- Air pollution caused by chronic traffic congestion and industrial emissions.

(c) (i) Reasons for urban sprawl might include:
- Growth of population.
- Growth of suburban housing both high quality private and low cost, council estates.
- Cheaper land prices on outskirts.
- Development of shopping malls, industrial estates and retail parks.
- Growth of leisure facilities eg golf courses, new football stadia.
- Need for motorway and by-pass developments.
- Increased level of commuting to suburban areas and villages with more attractive environments.
- Negative aspects of the city eg pollution, congestion, land prices, house prices, levels of social problems like crime.
- Increased number of single person households.

(ii) Candidates could look at problems at the edge of the city as well as within the inner urban areas. Problems might include:
- Urban sprawl using up recreational and farm land.
- Urban sprawl threatening wildlife habitats and removing clean air lungs and open land.
- Increased commuting leading to traffic congestion and increasing levels of air pollution.
- Buildings and services in inner urban areas not being used or becoming run down or derelict eg housing, schools, factories and shopping areas.

(iii) Candidates require to select one problem and identify how their chosen city has dealt with the problem. For example if the candidate had chosen traffic congestion the following solutions could be developed.
- Policies to reduce cars eg car sharing, high occupancy vehicle lanes, new car tax charges, congestion charges, cycle routes.

- Promotion of improved public transport, including lower pricing, integrated transit systems.
- Park and Ride schemes.
- Mass transit systems using fixed routes eg metro lines and tramways.
- Changing road systems eg flexi-time travel, tidal flows, coordinating traffic lights, bus lanes.

Question 5 – European Regional Inequalities

(a) (i) • Credit should be awarded for candidates noting that Convergence Regions are found in Europe's peripheral areas, notably in eastern periphery countries with former centrally planned economies eg Bulgaria, Hungary, Baltic states etc.
- Also southern European peripheral areas such as southern Italy and much of Greece, Spain and Portugal.
- In the UK, Cornwall and western Wales have this status.
- No Convergence Regions found in most of northern, central and western areas of the EU.

(ii) EU measures could include:
- Cohesion Fund – aimed at member states whose Gross National Product (GNP) per inhabitant is less than 90% of the EU average. It serves to reduce their economic and social shortfall, as well as to stabilise their economy. The Cohesion Fund finances activities such as trans-European transport networks and also projects related to energy or transport as long as they clearly present a benefit to the environment.
- The European Regional Development Fund (ERDF) aims to strengthen economic and social cohesion in the EU by correcting imbalances between its regions by financing technical assistance measures, improvements to local infrastructure etc.
- The task of the European Investment Bank (EIB), the EU's financing institution, is to contribute towards the integration, balance, development and economic and social cohesion of all the member states.
- The European Social Fund (ESF) set out to improve employment and job opportunities in the EU through lifelong learning schemes and providing access and employment for job seekers, the unemployed, women and migrants. It supports actions to socially integrate disadvantaged people, combating discrimination in the job market.

For full marks, candidates will require to illustrate points made with some well-chosen examples and statistics.

(b) (i) Candidates should use some form of comparative statements covering all four indicators to get full marks.

The four indicators given all identify a similar pattern identifying regional inequalities within the UK –East England, SE and SW England and London generally fare better than Wales, the West Midlands and areas further north.
- Population change – apart from Northern Ireland, regions with the highest increase in population are in the south of England. No growth or decrease in Scotland, NE and NW England.
- Average house prices – highest in London and SE England, while Scotland, northern regions of England and Wales have figures well below the UK average.
- Gross Disposable Household Income – very similar to house prices although Yorks and Humberside, East Midlands and Scotland fare slightly better.

- Working Age Population with no qualifications (%) – very low in southern England, again high in Wales, northern England and especially Northern Ireland. Scotland same as UK average.

(ii) The UK's regional inequalities stem from a combination of the physical differences between the higher and steeper land to the north and west of the UK compared with the lower and more gently sloping land to the south and east coupled with the remoteness of the north-west compared to the proximity of the south-east to the 'core' of the EU. Candidates may justifiably stress the positive and negative aspects of different regions.
- Physical factors might mention advantages/ problems such as relief, rock types, climate and water supply, soil fertility and erosion.
- Human factors might mention decline of traditional heavy industries, growth areas of new lighter industries and hi-tech industries, out-migration from north and differences in accessibility related to communications and remoteness.

(iii) UK national government help could include:
- UK government identifies 'Assisted Areas' eligible for regional selective assistance in line with EU moves to redistribute most of aid budget to poorer areas. Assisted areas include the whole of Northern Ireland, Cornwall and the Scilly Isles, West Wales and the Valleys and the Scottish Highlands and Islands.
- There are also economic development agencies in the four countries of the UK which aim to attract investment and help new and existing businesses compete nationally and internationally.
- The Welsh Assembly Government's department of Economy and Transport in Wales.
- Scottish Enterprise and Highlands and Islands Enterprise in Scotland.
- Invest Northern Ireland in NI and The Regional Development Agencies (RDAs) in England.

Question 6 – Development and Health

(a) Candidates should be able to refer to:
- Oil rich countries such as Saudi Arabia, Brunei; well-off countries like Malaysia which can export primary products such as hardwoods, rubber, palm oil and tin.
- Poor Sahelian countries like Mali, Chad and Burkina Faso which are landlocked, lack resources, have poor quality farmland, high levels of disease.
- Newly Industrialised Countries eg South Korea, Taiwan have high GNPs due to steel-making, shipbuilding, car manufacturing, clothing etc. Countries with entrepreneurial skills and low labour costs.
- Large countries eg Brazil with a variety of opportunities ranging from resources in Amazonia to tourism in the South East around Rio.
- Tourist destinations eg Sri Lanka, Thailand, Caribbean islands like Barbados, earn foreign currency and improve living standards and create new job opportunities.
- Countries which suffer natural disasters which restrict development and cause massive damage to infrastructure. Examples include drought in Ethiopia, floods/cyclones in Bangladesh, hurricanes in Caribbean and tsunamis in Indonesia, earthquakes in China.
- Mountainous countries eg Tibet, Afghanistan which restrict communications and farming.
- Areas of political instability which divert aid and resources away from areas of need. Examples include civil war in Sudan, large scale conflicts in Afghanistan and Iraq, corruption, mismanagement and need for regime change in Zimbabwe.

(b) Candidates can make use of resources to suggest the following factors which may lead to low life expectancy.
- Chad is landlocked restricting trade, income from imports/exports and reducing money available to improve quality of life.
- Low GDP means Chad will struggle to provide services such as hospitals/clean water/sanitation which will increase ill health.
- High infant mortality will reduce average life expectancy.
- Low quantities of farmland and irrigated land mean crop production will be less than required to feed population leading to malnutrition and ill health.
- Low literacy levels imply poor education in areas of hygiene/birth control/disease control.
- Inhospitable areas eg desert, uplands means living conditions are very harsh
- High fertility rates mean large families with not enough food and resources to go round.
- High levels of water borne diseases reduce life expectancy.

(c) (i) Answers will depend on the disease chosen but for Malaria might include:

Physical factors:
- Hot wet climates such as those experienced in the tropical rainforests or monsoon areas of the world.
- Temperatures of between 15°C and 40°C.
- Areas of shade in which the mosquito can digest human blood.
- Migration.
- Not completing course of drugs.

Human factors:
- Suitable breeding habitat for the female anopheles mosquito – areas of stagnant water such as reservoirs, ponds, irrigation channels.
- Nearby settlements to provide a 'blood reservoir'.
- Areas of bad sanitation, poor irrigation or drainage.
- Exposure of bare skin.
- Migration.
- Not completing course of drugs.

(ii) Measures taken to combat malaria may include:
Trying to eradicate the mosquitoes:
- Insecticides eg DDT.
- Newer insecticides such as Malathion
- Mustard seed 'bombing' – become wet and sticky and drag mosquito larvae under the water drowning them.
- Egg-white sprayed on water – suffocates larvae by clogging up their breathing tubes.
- BTI bacteria grown in coconuts. Fermented coconuts are, after a few days, broken open and thrown into mosquito-infested ponds. The larvae eat the bacteria and have their stomach linings destroyed!
- Larvae eating fish.
- Drainage of swamps.

Treating those suffering from malaria:
- Drugs like chloroquin, larium and malarone.
- Quinghaosu extract from the artemesian plant – a traditional Chinese cure.
- Continued search for a vaccine – not available as yet.
- Education programmes in –
  - the use of insect repellents eg Autan
  - covering the skin at dusk when the mosquitoes are most active
  - sleeping under an insecticide treated mosquito net
  - mesh coverings over windows/door openings.
- WHO 'Roll back malaria' campaign.
- The Bill and Melinda Gates Foundation.

(iii) The benefits of controlling the disease on a developing country might include:
- Saving money on health, medicines, doctors, drugs etc.
- Reduction in the national debt.
- Healthier workforce and increased productivity.
- Longer life expectancy and decreased infant mortality rates.
- Scarce financial resources could be spent on other areas such as education or housing.
- More tourists/foreign investment may be attracted if there was less risk of disease – leading to more job opportunities, foreign currency earnings, increased prosperity.

GEOGRAPHY HIGHER PHYSICAL AND HUMAN ENVIRONMENTS 2011

SECTION A

Question 1 – Atmosphere

(a) Maritime Tropical (mT)
- Origin – Atlantic ocean/Gulf of Guinea, in tropical latitudes
- Weather characteristics – hot, high humidity, warm
  Nature – unstable

Continental Tropical (cT)
- Origin – Sahara Desert, in tropical latitudes
- Weather characteristics – hot/very hot, dry, low humidity, warm
  Nature – stable, poor visibility

(b) Description should highlight the marked contrast in precipitation totals, seasonal distribution and number of days between a very dry north (Gao with only 200 mm in a hot desert climate in Mali) and a much wetter south (Abidjan with 1700 mm in a tropical rainforest climate in the Ivory Coast).

Bobo-Dioulasso in Burkina Faso in central West Africa has an 'in-between' amount of both rain days and total annual precipitation (1000 mm in a Savannah climate).

Candidates should also refer to the variation in rain days and seasonal distribution for each station. Gao with a limited amount of precipitation in summer, Bobo-Doiulasso with a clear wet season/dry season regime and Abidjan with a 'twin-peak' regime with a major peak in June and a smaller peak in October/November.

Explanation should focus on the role of the ITCZ and the movement of the Maritime Tropical and Continental Tropical air masses over the course of the year. For example, Abidjan, on the Gulf of Guinea coast, is influenced by hot, humid mT air for most of the year, accounting for its higher total annual precipitation and greater number of rain days. The twin precipitation peaks can be attributed to the ITCZ moving northwards in the early part of the year and then southwards later in the year in line with the thermal equator/overhead sun.

Gao, on the other hand, is under the influence of hot, dry cT air for most of the year and therefore has far fewer rain days and a very low total annual precipitation figure as it lies well to the north of the ITCZ for most of the year. Bobo-Dioulasso again is in an 'in-between' position, getting more rain days and heavy summer precipitation from June-August when the ITCZ is furthest north.

Question 2 – Biosphere

(a) Climax vegetation is the final stage in the development of the natural vegetation of a locality or region when the composition of the plant community is relatively stable and in equilibrium with the existing environmental conditions. This is normally determined by climate or soil. These are self-sustaining ecosystems.

Candidates should be credited for being able to demonstrate knowledge of the evolution of plant life from early colonisation by pioneer species then, by succession, to the ultimate vegetation climax. Appropriate examples could also be given credit eg oak-ash forest in a cool temperature climate such as exists over much of Britain or Scots pine-birch forest in colder, wetter and less fertile Highland environments.

(b) **Strandline (Sea Sandwort, Sea Rocket, Saltwort, Sea Twitch)**
These are all salt tolerant (halophytic) species and can withstand the desiccating effects of the sand and the wind. Some can even withstand periodic immersion in sea water. There is a high pH here (alkaline conditions) due to the presence of sea shells. The presence of these plants leads to further deposition of sand and the establishment of less hardy species.

Embryo Dune (Sea/Sand Couch, Lyme Grass, Frosted Orache, Sea Rocket)
These dune pioneer species grow side (lateral) roots and underground stems (rhizomes) which bind the sand together. These grassy plants can also tolerate occasional immersion in sea water. Some species on the strandline are also found in the embryo dunes.

Fore Dune (Sea Bindweed, Sea Holly, Sand Sedge, Marram Grass)
A slightly higher humus content (from decayed plants), and lower salt content (further from the sea) allows these species to further stabilise the dune and allow the establishment of Marram Grass which becomes a key plant in the build up of the dune.

Yellow Dune (Marram Grass, Sand Fescue, Sand Sedge, Sea Bindweed, Ragwort)
Both the humus content and the acidity of the soil have increased at this location. Marram can align itself with the prevailing wind and curl its leaves to reduce moisture loss; it can also survive being buried by the shifting sand of the dune. As sand deposition increases the Marram responds by more rapid rhizome growth (up to 1 metre a year). It is xerophytic, and so is better able to survive the dry conditions of the dune. It also has long roots which help to bind deposited sand and anchor it into the dunes as well as access water supplies some distance below. All these factors allow it to become the dominant species on the Yellow Dune.

Grey Dunes (Sand Sedge, Sand Fescue, Bird's Foot Trefoil, Heather, Sea Buckhorn, Grey Lichens eg Cladonia species)
As a result of an increase in organic content (humus), greater shelter and a damper soil, a wider range of plants can thrive here. Marram dies back (contributing humus) to be replaced by other grasses. As a result of leaching and the build up of humus the soil is considerably more acidic allowing more plant species to flourish.

Slacks (reeds, rushes, cotton grass, flag iris, alders and small willow trees)
In the wetter slacks, close to the water table, several water loving (hydrophytic) species may survive.

Climax (Heather, trees such as Birch, Pine or Spruce)
In some areas heathland may dominate with a range of heathers being prominent. Eventually trees such as Birch, Pine or Spruce could establish a hold. In the shell rich areas of the Western Isles, Machair may develop.

Question 3 – Rural Geography

(i) For **shifting cultivation** the main characteristics of the landscape might include:
- clearings are made in the rainforest by cutting down and burning trees
- largest trees and some fruit-bearing trees may be left for protection/food (some are too difficult to remove)
- the 'cultivation' part refers to the practice of growing crops (manioc/cassava, yams …) in the clearing, using ash from the tree burning as fertiliser

- the 'shifting' part refers to the practice of moving to another clearing as the soil becomes exhausted and crop yields fall
- low population density due to large area of land needed
- settlements could be fixed (and rotational clearings made around them) or the housing may also be abandoned and left to biodegrade before the tribe returns to the area.

(ii) For **shifting cultivation** the main changes might include:
- loss of traditional tribal land due to cattle ranching, mineral extraction, logging, HEP development
- change in land use with set reservations/settlements, National Parks and conservation areas
- climate change with increasing unpredictability of drought/flood cycles.

For **shifting cultivation** the main impacts might include:
- population movement into inaccessible areas which are often less fertile
- rural depopulation with shanty town growth in larger urban areas
- contact with Western culture can bring diseases, alcohol/drug misuse
- population densities increase in remaining areas, putting more strain on limited land and a shorter fallow period.
- decreasing soil fertility and output per hectare
- soil erosion can take place with the soil choking the rivers reducing fish/wildlife in the area/impact on diet.
- pollution from other land users ie mercury used in gold extraction can impact on the health of the locals
- the impact of global warming on biodiversity and medicinal cures.

Question 4 – Industry

(a) Reasons for location of industry in Swansea may include:

Physical factors:
- Area A on the outskirts of town with flat land, and room for expansion, Area B is on flat floodplain of Afon Tawe.
- Both have flat land for easy construction of industrial buildings.

Human factors:
- Proximity to local market in South Wales.
- Access to docklands for import and export and via Afon Tawe.
- Close to motorways for easy access of materials and finished goods,a and branch line 683970
- Proximity to local labour force.
- Edge of town – cheaper land.
- Close to universities for skilled graduates and research facilities.
- Close to other modern industries that may supply components or share resources.
- Swansea airport for visiting executives, or transporting light products (5691).
- Pleasant working environment.

(b) Answers will vary, depending on industrial concentration chosen.

Explanations may include:
- Creation of Enterprise Zones.
- Welsh Development Agency.
- Creation of new town (Cwmbran).
- Rent free accommodation.
- Grants.
- Retraining schemes.
- Relocation of industry (government offices eg DVLA, and foreign businesses eg Sony).
- Tax incentives.

- Road and infrastructure projects (M4).
- Environmental improvement schemes.
- Objective 1 funding.

SECTION B
Question 5 – Lithosphere

(a) Descriptions could include:
- Cliffs eg Newton Cliff (GR 600870).
- Headland eg Pwlldu Head (GR 570863).
- Caves eg Mitchin Hole Cave (GR 555869).
- Blow Holes eg Bacon Hole (GR 561868).
- Shore (wave-cut) platform (GR 615869).
- Stack eg Mumbles Head (GR 636871).
- Skerries (stack or stump) eg Rothers Sker (GR 612869).
- Bay (GR592874) – Caswell Bay

(b) **A quality diagram could achieve full marks.**
Candidates should refer to the processes of coastal erosion ie hydraulic action, abrasion, solution and attrition. A typical answer may include:
Caves are most likely to occur where the coastline consists of hard rock and is attacked by prolonged wave attack along a line of weakness such as a joint or fault. The waves attack the weakness by abrasion, hydraulic action or solution. Over time, horizontal erosion of the cave may cut through the headland to form an arch. Continued erosion of the foot of the area may eventually cause the roof to collapse leaving a stack, isolated from the cliff.

Question 6 – Hydrosphere

(a) Answers should refer to the four elements in a drainage basin:
- input: precipitation
- storage: surface storage eg lakes, soil moisture, ground water, interception
- transfers: surface run off eg tributaries, throughflow, groundwater flow, infiltration, throughfall, percolation, stem flow
- outputs: transpiration, evaporation, surface run-off (rivers)

(b) Answers should identify various parts of the river level graph.
- Steady river level (under 0.4m) until 03:00 hours due to an initial lack of rain and then small amounts of rain at 05:00 hours (0.5mm) and 06:00 hours (0.8mm) infiltrate the soil (after interception by vegetation) and the river level starts to increase slowly.
- The river level continues to rise at a steady rate from 07:00 hours to 10:00 hours, due to the increase in rainfall totals and duration. The heavier rain is filling up storages in the soil because of throughflow and groundwater. The soil is now saturated, so water runs off the land and enters the river quickly leading to a potential flood situation.
- The peak rainfall occurs at 08:00 hours (6.2mm) and the peak river level occurs at 18:00 hours (0.7m). This is a basin lag time of approximately 10 hours. This could be accounted for by vegetation cover, or by reference to geology or soil infiltration rates.
- From 14:00 hours to the end of the graph the rainfall declines and stops at 18:00 hours. The recession limb falls back towards base level as the supply of water is reduced.

SECTION C
Question 7 – Urban Geography

(a) Answers will vary according to the city studied but may include reference to:
Site
- Flat land.

- Inside a large river meander.
- Early functions eg religious, defensive, trading site.
- Raw materials.
- Lowest bridging point.

Situation
- Easily accessible to major settlements.
- Accessible to ports.
- Major route focus.
- Accessible to airports.

(b) Answers will vary according to the city studied but for the inner city changes might include:

Description of the changes:
- Population reduced as people moved out of the area.
- Redevelopment of housing and the area.
- Construction of council houses and flats on cleared areas of the old inner city.
- Demolition of some terraces, improvements to others (indoor toilets etc).
- Environmental improvements eg parks, community centres, leisure centres.
- Relocation of industry and the closure of industry.
- Changes in the ethnicity of residents.

Explanation of the changes:
- Area was overcrowded.
- Housing was in a very poor condition and amenities were poor.
- High levels of unemployment, poverty and social problems such as crime.
- The environment was very poor and inward investment was difficult to attract.
- Industry declined as manufacturing moved to countries with lower labour costs.
- Industry has moved to modernised areas on the edge of the city in new custom-built units.
- Congested traffic and air pollution put off potential investors.
- Movement out to the suburbs due to the desire for a better quality environment.

Question 8 – Population Geography

(a) Between censuses there is compulsory registration of births, deaths and marriages by the General Register Office for each part of the UK. The other main changes are brought about by emigration and immigration and the Home Office's UK Border Agency records migration into the UK.

Also mini or sample censuses are carried out such as the 2009 Census Rehearsal for England and Wales as well as government sponsored sample surveys of population and social trends.
The UK official 2011 Census website identifies 6 areas of accurate population data for the targeting of taxpayers' money.
- Population numbers – to calculate grants for local authorities to plan eg schools and teacher numbers.
- Health – to know the age and socio-economic make-up of the population to allocate health and social services resources.
- Housing – to ascertain the need for new housing.
- Employment – to help government and businesses plan jobs and training policies.
- Transport – to identify where there is pressure on transport systems and for planning of roads and public transport.
- Ethnic Group – to identify the extent and nature of disadvantage in Britain.

(b) Difficulties affecting accurate population data collection in Developing Countries might include:

- Countries suffering from a continuing war situation such as Afghanistan.
- Illegal immigrants wishing to avoid detection eg Mexicans in southern California.
- The cost involved in carrying out a census is prohibitive to many Developing Countries – training enumerators, printing and distributing forms etc.
- The sheer size of some developing countries eg Indonesia with many islands spread over a large area.
- Suspicion of the use of data collected, eg China's one-child policy with many female births unrecorded.
- Countries with large numbers of migrants eg rural-urban migration into massive shanty towns eg Kibera in Nairobi, refugees from Rwanda in Burundi etc.
- Nomadic people such as the Tuareg in West Africa, shifting cultivators in Amazonia.
- Poor communication links eg mountain regions of Bolivia.
- Low levels of literacy and variety of languages spoken within a country eg India has 15 official languages.

GEOGRAPHY HIGHER ENVIRONMENTAL INTERACTIONS 2011

Question 1 – Rural Land Resources

(a) For an answer to achieve full marks, well annotated diagrams must be used. For full marks a minimum of two features must be described and explained, eg for a corrie points could include:
- Snow accumulates in north/east-facing hollow due to lack of melting.
- Successive layers of snow compress into ice/neve.
- Ice moves downhill under gravity.
- Freeze-thaw weathering occurs on backwall.
- Plucking steepens the backwall.
- Boulders embedded in ice grind away at bottom of the corrie.
- Abrasion carves out armchair-shaped depression due to rotational movement.
- Rate of erosion decreases at edge of corrie leaving a rock lip.

(b) Answers are expected to link these opportunities to the physical landscape, and answers must mention both social and economic opportunities for full marks.

Explanations can be developed from:

Social opportunities
- Mountaineering and hillwalking.
- Forest walks, picnic sites and orienteering courses.
- Sailing, fishing and other water sports.
- Nature conservation.

Economic opportunities
- Tourism and associated employment and profits.
- Development of hotels, bunkhouses and campsites.
- Hill sheep farming.
- Forestry plantations.
- HEP and water supply.
- Quarrying.

(c) Answers should be able to compare the popularity of parks based on analysis of the resource:
- Lake District is close to heavily populated areas, Merseyside, Yorkshire, Manchester.
- Snowdonia more remote in North-West Wales.
- Lake District is more accessible by motorway especially for short visits, M6, M74.
- Snowdonia is less accessible by motorway.
- Credit can also be given for candidates' knowledge of both parks, in terms of attractions.

(d) (i) Answers should be able to explain the environmental conflicts including:
- Traffic congestion especially on narrow rural roads and in car parks especially at peak holiday periods.
- Increased air and noise pollution.
- Increased holiday homes which leave rural areas empty during the week or off peak.
- Footpath erosion.
- Disruption to farms, damage to walls, disturbance to animals.
- Litter.
- Unsightly buildings including hotels, leisure complexes, caravan sites.
- Impact on lakes, bank erosion and diesel pollution due to water sports.

(ii) • One way streets, bypasses, wardens, parking restrictions.
- Encourage use of public transport eg park and ride, minibus.
- Use of cycle paths, bridle ways, long distance paths.
- Use of permits to separate locals and visitors.

Question 2 – Rural Land Degradation

(a) The four main processes of erosion by water can be described as:
- Rainsplash – the impact of raindrops on the surface of a soil.
- Sheet wash – the removal of a thin layer of surface soil which has already been disturbed by rainsplash.
- Rill erosion – small eroded channels, only a few centimetres deep and not permanent features, often obliterated by the next rainstorm.
- Gully erosion – steep sided water channels, several metres deep which can cut deeply into the soil after storms and are often permanent.

The three main processes of wind erosion can be described as:
- Surface creep – the slow movement of larger (and heavier) particles across the land surface.
- Saltation – the bouncing along of lighter particles.
- Suspension – the lightest particles (dust) blown off ground for up to several hundred kilometres, dust storms.

(b) Answers should include the following descriptions and explanations from the resources:
- Niger is a landlocked, dry country – part of the Sahel Zone.
- Extreme range of temperatures on a daily basis.
- Seasonal rainfall concentrated from April to September.
- High temperatures coinciding with highest precipitation leading to high evaporation.
- Annual rainfall creates desert conditions.
- Variable annual rainfall from 1950 to 2010 with periods above average encouraging farming even in marginal areas, and periods below average leading to drought and degradation.
- Clear skies and strong direct sun can bake ground.

Candidate must relate aspects of climate to degradation caused by wind and water erosion.

(c) Answers should include explanations in the four areas of human activity outlined (Africa north of the Equator):
- Deforestation for firewood and farmland left soil exposed to erosion, removed root systems which would hold soil together, and removed shelter belts and wind breaks.
- Overgrazing exposed soil to winds by loss of vegetation cover, hooves break up soil making it susceptible to wind and water erosion, and in some cases compact the soil, especially near water holes.
- Overcropping means soil structure breaks up, with monoculture depleting nutrients, reduced fallow times meaning soil cannot rest or recover, marginal land eg slopes being used and becoming susceptible to wind and water erosion.
- Inappropriate farming techniques including monoculture, inappropriate ploughing eg deep ploughing of fragile soils, irrigation leading to salinisation, lack of organic fertilisers used.

(d) Soil conservation strategies might include:
- Crop rotation.
- Diversification of farming types.
- Keeping land under grass or fallow.
- Trash farming/stubble mulching.
- Replanting shelter belts.
- Strip cultivation and intercropping.
- Improved irrigation.
- Soil banks.

- Contour ploughing.
- Terracing.
- Use of natural fertilisers.
- Gully repair.
- Re-afforestation of slopes and marginal land.

Question 3 – River Basin Management

(a) Candidates may mention a range of reasons to explain the need for water management including:
- Very high rainfall in Borneo (tropical rainforest conditions).
- Flood control.
- Regulating flow and storage of water.
- Power supply for expanding cities and industry.
- Export of surplus electricity.
- Water for industrial purposes.
- Drinking water for increasing population.

(b) Physical factors might include:
- Solid foundations for a dam.
- Consideration of earthquake zones/fault lines.
- Narrow cross-section to reduce dam length.
- Large, deep valley to flood behind the dam.
- Lack of permeability in rock below reservoir.
- Sufficient water supply from catchment area.
- Low evaporation rates.
- Impact on the hydrological cycle.

(c) (i) Answers should be authentic for the chosen river basin. Answers will depend upon the basin chosen. However, some suggestions are outlined below:

Social:
- Greater population can be sustained with increased fresh water supply.
- Less disease and poor health due to better water supply/sanitation.
- Recreational opportunities/tourism on rivers and reservoirs.
- More widespread availability of electricity (and therefore modern technology/development).
- Floods could be avoided.

Economic:
- Improved farming outputs with possible surplus for sale.
- HEP – industrial development creating job opportunities eg Borneo could export surplus electricity to Indonesia and transfer to mainland Malaysia.
- Water (and power) for industry eg a proposed aluminium smelter in Borneo linked to the Bakun Dam.
- Navigation opportunities.

Environmental:
- Increased fresh water supply improves sanitation and health.
- Scenic improvement.

(ii) Answers will depend on the dam studied but for the Bakun Dam, answers may include:
- Threatened wildlife – endangered species may vanish, loss of tourism revenue, loss of medicinal plants/drugs still to be discovered.
- Forest cover – deforestation from reservoir/dam construction and increased access to remote areas with the resulting impact on local climate and global warming, silting up of rivers/soil erosion.
- Changing landscapes – reduced supply of timber for world trade and the increased planting of palm oil for biofuels/oil exports replacing the rainforest and resultant loss of biodiversity as access improves.
- People – relocation of indigenous tribes, loss of land/traditional way of life, spread of water-borne diseases from reservoirs.

The candidate may also include facts from the reference diagrams:
- There is already a surplus of electricity.
- The transmission line to transfer more electricity to the Malaysian mainland is only 'proposed'.

Other factors such as cost or political factors could be made relevant by the candidate.

Question 4 – Urban Change and its Management

(a) Answers will depend on the Developed World city chosen, but for the UK answers might suggest favourable locations for cities eg:
- Coastal locations – for trade with Europe/America (London/Glasgow), fishing industry (Aberdeen) and ship-building industry (Glasgow/Belfast).
- Natural routeways/rivers/canals – for communication, trade in raw materials (Leeds, Manchester, Sheffield, Birmingham).
- Access to raw materials – for coal, iron ore and limestone for the iron and steel industry (Glasgow/Sheffield).
- Historical/political factors in location of capital/primate cities – royal residences, parliaments etc (London, Edinburgh, Cardiff).

Whereas negative locations may also be included eg:
- Mountainous/upland areas – Highlands, Pennines.
- Inaccessible/marshy areas – Islands, Fens

(b) Candidates should note the advantages for the residents of the East End of Glasgow using the statistics from the diagrams/table.

Advantages could include:
- Improved communications – motorway/road extensions and improvements.
- Jobs – in the various venues before, during and after the event (construction, catering, transport etc).
- Training – the promise of skills training in various volunteer roles during the competitions that could then lead to permanent jobs.
- Social – attending/participating in the competitions.

Disadvantages could include:
- Disruption and pollution from the construction process in producing the 30% of the venues/facilities still to be built.
- Loss of land, access, community spirit/cohesion that new roads/motorways can lead to, particularly with the M74 extension cutting through a densely populated area.
- Some landowners/developers may have had to sell their property under compulsory purchase orders that may have led them to lose potential profits.

(c) Traffic congestion in a Developed World city. For Glasgow, candidates might suggest:
- An urban core developed in the pre-car era with medieval/Victorian sections unsuited for modern traffic (narrow, cobbled, grid-iron pattern with many junctions).
- Increased commuting from dormitory towns and villages converging on a few main arteries (Paisley Road West, Great Western Road, M8, Kilmarnock/Ayr Road).
- Major roads converging to cross the River Clyde (Clyde Tunnel, Kingston Bridge).
- Glasgow is a growing tourist/shopping centre attracting coach tours and shoppers from a larger hinterland.
- More stringent traffic regulations in and around the CBD with a shortage of cheaper gap site car parking facilities.
- Growing car ownership and school run traffic extending and expanding the 'rush hours'.
- Increased road haulage/deliveries in larger vehicles with a resulting need for more road maintenance.

(*d*) (i) The problems should be relevant to the candidate's chosen city and might include:
- Chaotic urban infrastructure eg incomplete water and sewerage supplies and connections leading to the spread of disease.
- Unemployment/underemployment
  – Growth of the 'grey' economy and black market
  – Drugs, crime, racketeering and prostitution are common and often involve a greater % of the population than a city in developed country
  – Poor wages for unskilled jobs partly due to the huge supply of labour available.
- Lack of services, schools and hospitals.
- Difficulties in encouraging city/public employees to work in the 'shanty' areas.
- Chronic traffic congestion and associated high levels of atmospheric pollution
  – Proliferation of 'informal' city transport (having both advantages and disadvantages.)
- Continued growth of 'shanty towns' in a range of locations in and around the city
  – 'Natural' disasters such as landslides resulting from the inappropriate building techniques and methods on fragile or unsafe land.

(ii) Again the methods used to tackle the problems should be related to the candidate's chosen city.
Candidates could offer a number of 'generic' solutions:
- An increase in the empowerment of the local people often with the aid of charity/church groups which provide advice/counsel/ lobbying facilities for the poorest elements of the population.
- Local council plans to improve basic infrastructure, including provision of water/sewerage to established 'shanties'.
- Improvements in the standard of basic education.
- The provision of hardware/utilities with the local populace providing the skill/effort to install these ie the 'basic shell' of housing being provided.

Specific solutions related to the candidate's chosen city are also wanted, these could include government drives to demolish squatter settlements and re-house the residents in new housing schemes.

Question 5 – European Regional Inequalities

(*a*) Countries may wish to become members of the European Union for the following reasons:
- Removing trade barriers to boost growth and create jobs.
- Tackling climate change and promoting energy security.
- Improving standards and rights for consumers.
- Fighting international crime and illegal immigration.
- Bringing peace and stability to Europe by working with its neighbours.
- Giving Europe a more powerful voice in the world.
- Securing food supplies and essential raw materials.
- Improving standards of living in the member states.

Specific EU measures to aid development include:
- European Regional Development Fund (ERDF) which provides a wide range of direct and indirect assistance to encourage firms to move to disadvantaged areas eg loans, grants, infrastructure improvements.
- European Investment Bank (EIB) provided loans for businesses setting up in disadvantaged areas.
- European Social Fund (ESF) assists with job retraining and relocating.

(*b*) Evidence might include the marked differences in GNP per capita between the "North" and the "South". Figures should be quoted and regions named, Abruzzi, Molise, Sardinia, Campania and Sicily stand out as being particularly

disadvantaged compared to regions such as Emilio Romagna or Lombardy in the "North". Regions such as Umbria and Latium could be said to be in a middle category or transitional. The concentration of industry, commerce and services in the North should be noted.

(*c*) Answers will be dependent on the country chosen. For Italy, the following factors may be mentioned:

Physical factors
- Relief/geology.
- Climate/water/resources.
- Soil quality/soil erosion.
- Natural disasters.

Human factors
- Remoteness/isolation/communications.
- Limited employment opportunities in the South.
- Decline of traditional industry.
- Land tenure problems.
- Unskilled labour, poorly educated workforce.

(*d*) (i) Once again answers will depend upon the country chosen.

National Government Measures include:
- Regional development status, Enterprise Zone status, capital allowances, training grants, assistance with labour costs.
- Specific assistance to former coal mining/iron and steel areas.
- Intervention of national government resulting in the relocation of major government employers or state owned firms to disadvantaged areas eg Fiat to Southern Italy, DVLA in Swansea, MOD in Glasgow.
- In Italy the Cassa il Mezzogiorno would be a key policy.
- Tesco Finance to Glasgow - £5 million Regional Selective Assistance (RSA) grant.

(ii) Comment should be made on the effectiveness of the measures outlined eg the long term benefits or disadvantages of using these incentives.

Question 6 – Development and Health

(*a*) (i) Candidates should be able to identify several differences between provinces using the figures provided. It is clear that the North Eastern province is by far the least developed across the development indicators and that Central province is clearly at the highest level of development within Kenya.

Candidates should get credit for noting that the table covers the three major areas of education, wealth and health and could comment on each of these in turn.

- Education – varies from 87% of females with no education in North Eastern province to only 10% in Nairobi. The striking difference between male and female percentages, especially in poorer provinces, with males getting preferential treatment in many developing countries, could be noted.
- Wealth – all areas of Kenya have many poor, but again big variation from almost 2/3 in North Eastern, Western and Nyanza to <1/3 in Central.
- Health – huge variation again here with >3/4 of children in Central province having all vaccinations whereas only 8% in North Eastern are protected. The North Eastern province trails all others alarmingly in this indicator.

(ii) Answers will, obviously, depend on the Developing World country chosen but for Brazil could include:

- The South East is much more prosperous than other regions due to the concentration of industry and commerce in the "Golden Triangle" of Sao Paulo, Rio de Janeiro and Belo Horizonte. This area has the best transport system in Brazil, the greatest number of services, and has benefited most from Government help. Coffee growing has long been carried out on the rich terra rossa soils around Sao Paulo producing job opportunities and creating wealth for the area and the national economy. Rio de Janeiro – until 1960 the capital of Brazil, had the advantages of a good natural harbour which encouraged trade, immigration, industry, and more recently, tourism.
- The North East, in contrast, is handicapped by more negative factors such as periodic droughts, fewer mineral resources and a shortage of energy supplies all of which have encouraged outwards migration.
- The North (Amazonia) suffers from its more peripheral location, its inhospitable (rainforest) climate, poor soils, dense vegetation and inaccessibility. Not surprisingly, it is the poorest of Brazil's five main regions. Until recently, there was a lack of Government investment and much of the region lost out on basic services such as health, education and electricity.
- In addition to explaining the sorts of marked socio-economic regional variations which exist in a huge and diverse country such as Brazil, candidates may also comment on the marked differences in living standards which exist between relatively wealthy and better-provided-for urban areas compared to poorer more isolated rural areas and to the contrasts that can be found *within* urban areas – eg hillside favelas such as Rocinho in Rio versus the prosperous apartments overlooking Copacabana Beach.

(b) Measures used to combat the spread of malaria can include:

Trying to eradicate the mosquito/mosquito larvae:
- Pesticides/insecticides such as DDT and later Malathion.
- Mustard seeds thrown on water areas become wet and sticky and drag the mosquito larvae under the water, drowning them.
- Egg-white sprayed on water surfaces creates a film which suffocates the larvae by clogging up their breathing tubes.
- BTI bacteria grown in coconuts – the fermented coconuts are broken open after a few days and thrown into the mosquito larvae-infested ponds. The larvae eat the bacteria and have their stomach lining destroyed.
- Putting larvae-eating fish such as the muddy loach into ponds.
- Draining swamps, planting eucalyptus trees which soak up excess moisture, covering standing water.
- Genetic engineering, eg engineering sterile male mosquitoes.

Treating those suffering from malaria:
- Drugs like quinine, chloroquine, larium and malarone.
- Quinghauso extracted from the artemesian plant – a traditional Chinese cure.
- Continued search for a vaccine – not available as yet.
- The WHO 'Roll Back Malaria' campaign.
- Research carried out by the Bill and Melinda Gates Foundation.
- Education programmes in:
  - the use of insect repellents such as Autan
  - covering the skin at dusk and dawn when the mosquitoes are most active
  - sleeping under an insecticide-treated mosquito net
  - mesh coverings over windows/door openings.

(c) Primary Health Care (PHC) strategies may include:
- Use of barefoot doctors ie trusted local people who can carry out treatment for certain common illnesses, often using cheaper, traditional remedies.
- Use of Oral Rehydration Therapy (ORT) to tackle diarrhoea and dehydration which can kill babies and young children.
- Vaccination programmes against diseases such as polio, measles and cholera. PHC can focus on preventative rather than more expensive curative medicine.
- Health education schemes in schools and communities, targeting children and women in relation to hygiene and diet. Using songs, posters, word of mouth rather than written information in societies with high illiteracy, especially among women.
- Local initiatives backed up by small local health centres staffed by doctors who can refer more serious/complex cases to hospitals.
- PHC can also be involved in provision of clean water supply eg WaterAid's work in Tanzania. Also construction of pit latrines/Blair toilets for decent sanitation, often with community participation.

GEOGRAPHY HIGHER
PHYSICAL AND HUMAN ENVIRONMENTS
2012

SECTION A

Question 1 – Lithosphere

(a) Evidence which suggests that Area A on Map Q1 is a glacial erosion landscape could include:
- Corrie and tarn – Glaslyn 6154, Llyn Coch 5954 (Llyn = lake)
- Corrie – 6055
- Lyn Du r Addu
- Hanging valley and waterfall – Cwm Llan 6152 with waterfall at 623517, Afon Merch and waterfalls in 6352 (Afon = river)
- Ribbon Lake – Llyn Gwynant 6451 and 6452
- Glacial Trough/U shaped Valley and misfit stream – Afon Glaslyn 6552 and 6553
- Pyramidal Peak – Snowdon summit 610544
- Arête – Crib Goch 621553, Crib y Ddysgl 615552 (Crib = ridge) – Bwlch Main 605538, Bwlchysaethau 616542 (Bwlch = mountain pass/gap)
- Truncated Spur – 650536, 645527
  (Craig = rock and Pen – y = the head of a valley)

(b) A sequence of diagrams, fully annotated, could score full marks.

In explaining the formation of terminal moraine, for example, candidates could refer to such points as:
- Moraine is material transported by a glacier.
- When the glacier reaches lower altitudes (or temperatures rise) the ice melts and deposits the moraine at its snout.
- Terminal moraine marks the furthest point that a glacier reaches.
- It forms a jumbled mass of unsorted material that stretches across the valley floor.
- Once the ice has retreated, the terminal (or end) moraine can often form a natural dam, creating a ribbon lake.

In explaining the formation of a drumlin, candidates may refer to points such as:
- Drumlins are elongated hills of glacial deposits.
- They are formed when the ice is still moving.
- The steep 'stoss' slope faces upstream and the 'lee' is the more gentle, longer axis of the drumlin which indicates the direction in which the glacier was moving.
- The drumlin would have been deposited when the glacier became overloaded with sediment.
- As the glacier lost power, material was deposited, in the same way that a river overloaded with sediment deposits the excess material.
- The glacier may have experienced a reduction in power due to melting.
- If there is a small obstacle on the ground, this may act as a trigger point and till will build up around it.
- It may also have been reshaped by further ice movements after it was deposited.

In explaining the formation of an esker candidates may refer to points such as:
- Eskers are produced as a result of running water in, on or under the glacier.
- They are linear mounds of sand and gravel that commonly snake their way across the landscape.
- As the glacier melts, sub-glacial streams flow and deposit their load.

- When the glacier retreats the sediment that had been deposited in the channel is lowered to the land surface where it forms a linear mound, or hill, that is roughly parallel to the path of the original glacial river.
- Eskers consist of sorted materials, largest first.

Question 2 – Hydrosphere

(a) For **deforestation** candidates could describe how cutting down trees increases run-off, decreases evapo-transpiration (and therefore cloud formation) and leads to more extreme river flows as water is not intercepted and stored by the trees.

For **irrigation** candidates could describe how taking water from a river or underground store can reduce river flow, lower water tables and increase evaporation/evapo-transpiration by placing water in surface stores (ditches/canals) or by crops removing water from the cycle as they grow.

For **urbanisation** candidates could describe how removal of natural vegetation and replacement with impermeable surfaces and drains can speed up overland flow and evaporation and can lead to higher river levels. It also decreases the amount of water which returns to groundwater storage, possibly reducing the water table.

For **mining** candidates may refer to the silting up of lakes, rivers and reservoirs leading to reduced storage capacity in these areas. Mining may also lead to reduced vegetation cover leading to increased run-off, higher evapo-transpiration and cloud formation altering the rainfall pattern.

(b) A sequence of diagrams, fully annotated could score full marks.

In explaining the formation of a **floodplain and natural levee** candidates could refer to such points as:
- When a river floods it deposits material (the load) on its flood plain.
- As the water loses energy on leaving the river channel material is deposited in order from heaviest particles nearest the channel to lightest further out.
- A natural embankment is therefore built up in layers each time the river floods.
- Material is also deposited on the river bed as water breaks through the levee in times of flood.
- Some river beds and their levees can rise many metres above the flood plain over time as the load on the river bed and levees build up, exacerbating flooding when it occurs.

Ox-bow Lake
- As the outer banks of a meander continue to be eroded laterally through processes such as hydraulic action the neck of the meander becomes narrower.
- Eventually due to the narrowing of the neck, the two outer bends meet and the river cuts through the neck of the meander. The water now takes its shortest route rather than flowing around the bend.
- Deposition gradually seals off the old meander bend forming a new straighter river channel. Due to deposition the old meander bend is left isolated from the main channel as an ox-bow lake.
- Over time this feature may fill up with sediment and may gradually dry up.

Delta
- When the river flows into a calmer body of water – a sea or lake, it is forced to slow down and there is a resultant drop in energy.
- This causes the river to deposit its suspended material.
- The river channel flowing into the sea may divide into a number of channels called distributaries as alluvium is built up in the channel.

- The coarsest materials are deposited first as foreset beds and fine sediment as bottomset beds further out to sea.
- Over many years this material builds up to form a body of land known as a delta.
- Typically, deltas are shaped like the triangular Greek letter after which they are named.
- However, this shape is created only when the material is deposited uniformly over the whole area. If the particles are dropped at different rates a bird's foot shape is formed.
- Where tidal currents are strong deltas may not develop and the sediment is carried further out to sea.

Question 3 – Population

(a) **Changes**

Stage 1
- Total population fluctuates but population growth is low, as high Death Rate (DR) due to wars, famine and epidemics is balanced by high Birth Rate (BR) due to high infant mortality rate and lack of contraception.

Stage 2
- Rapid population growth as DR falls due to medical advances eg vaccinations, improved water supply and sanitation and marked decrease in Infant Mortality Rate (IMR).
- BR remains high due to lack of contraception and family planning, children seen as an 'economic asset' and parents wanting many children as an 'insurance policy' for being looked after in old age until IMR is seen to fall.

Stage 3
- Despite rapidly falling BR, continued rapid population growth as DR continues to fall, with continued improvements in medicine and standards of living.
- BR falls due to the awareness of family planning and that smaller families are needed with decrease in IMR; children now seen as an 'economic liability'.
- Population growth levels off at end of stage 3 as BR and DR reach similar low levels.

(b) Possible changes to the population include:
- Stage 5 of the DTM, falling birth rate and slightly higher death rate due to larger proportion of older people in the population.
- Declining population may occur.
- If migration continues there may be a more youthful population.
- 6% increase in the working age proportion up to 63% population in 2012.
- 4% increase in the population of pensioners to 20% of the population.
- 17% children (down 10%).

Issues for government may include:
- Need to maintain an active population large enough to allow levels of taxation to remain constant or raise retirement age.
- Need to ensure there are no future shortages in workforce – need to recruit immigrant labour/ease access for asylum seekers. This can lead to civil unrest/ethnic tension.
- Need to sustain demand for particular products or services eg schools, maternity hospitals, which if affected could lead to higher levels of unemployment.
- Ageing population gives increased cost of pension provision and unpopular decisions for government about how pensions should be funded.

Question 4 – Industrial Geography

(a) Reasons for industrial decline may include:
- Lack of local raw materials.
- Increased competition from overseas.
- Cheaper labour from competitors.
- Old fashioned/dated equipment.
- Increasing cost of transporting new materials and finished goods.
- Poor infrastructure of road and rail.
- EU and government grants/incentives running out.
- Rationalisation of foreign companies leading to overseas plants in EU closing.
- Falling demand as new products take over market.
- Restricted/dated working practices.

(b) The impact of industrial closures may include:
- Unemployment.
- Rise in cases of depression.
- Rise in crime rates.
- Closure of local schools.
- Associated service and supply industries close.
- Workers and their families migrate from the area.
- Shops close.
- Lack of investment and inflow of new industry due to ethos of decline.
- Factories and surrounding areas become derelict.
- Houses and closed shops are boarded up.
- Area looks rundown.
- Less pollution from older industries.
- Areas may attract government intervention for regeneration.
- Brownfield sites are cheaper for regeneration.
- Contamination of industrial sites means it can be expensive to reclaim for other uses.

Section B
Question 5 – Biosphere

A fully annotated diagram could achieve full marks for A and for B.

(a) The following characteristics could be described for a gley soil:
- Horizons – well defined Ao, A and B horizons.
- Colour – A horizon, dark brown/grey colour – B horizon, blue-grey with red mottling (iron compounds).
- Soil biota – lack of soil biota.
- Texture – A – silty, B – clayey – angular rocks frost heaved up into B horizon.
- Drainage – waterlogged, giving anaerobic conditions.
- Short roots of grasses/shrubs.

(b) The following features could be included for a brown earth soil:
- Natural Vegetation – deciduous forest vegetation provides deep leaf litter, which is broken down rapidly in mild/warm climate. Trees have roots which penetrate deep into the soil, ensuring the recycling of minerals back to the vegetation.
- Soil Organisms – soil biota break down leaf litter producing mildly acidic mull humus. They also ensure the mixing of the soil, aerating it and preventing the formation of distinct layers within the soil.
- Climate/Relief and Drainage – precipitation slightly exceeds evaporation, giving downward leaching of the most soluble minerals and the possibility of an iron pan forming, impeding drainage. Soil colour varies from black humus to dark brown in A horizon to lighter brown in B horizon where humus content is less obvious. Texture is loamy and well-aerated in the A horizon but lighter in the B horizon.

Question 6 – Atmosphere

(a) Explanations for the differences between tropical areas and polar areas may include:
- Sun's rays concentrated on tropical latitudes where rays strike vertically.
- Rays have less atmosphere to pass through at the Tropics so less energy is lost through absorption and reflection.
- Sun's angle in the sky decreases towards the Poles due to the earth's curvature which spreads heat energy over a larger area.
- Albedo differs between Tropics and Poles – darker forest surfaces absorb radiation and ice covered areas reflect radiation.
- Sun is higher in the sky between the Tropics throughout the year, focussing energy.
- No solar insolation at the winter solstices at the Poles.

(b) Descriptions of possible consequences may include:
- Rise in sea level.
- More extreme weather (and more variable) including floods, droughts, hurricanes, tornadoes etc.
- Extension and retreat of vegetation by altitude and latitude.
- Melting of ice sheets/icebergs.
- Impact on wildlife eg extinction of species.
- Increase in diseases eg malaria.
- Change in length of growing season.
- Some areas will become wetter, others drier.
- Changes to ocean current circulation.
- Changes in atmospheric patterns linking to monsoon, El Nino, La Nina etc.

Section C

Question 7 – Rural Geography

(a) Main features of the shifting cultivation system might include:
- Clearings are made in the rainforest by cutting down and burning trees.
- Ash is used as natural fertiliser.
- Some trees are left for protection from erosion or food (fruits and nuts).
- 'Shifting' part refers to the practice of moving to another clearing as the soil becomes exhausted quickly by heavy rains and lack of fertilisers. Land area required is large as cultivators move from area to area within forest.
- 'cultivation' part refers to the practice of growing crops in the clearing such as manioc/yams/cassava.
- System is labour intensive with small labour force due to subsistence nature of system which is unable to support a large population.
- Very low input of capital related to subsistence nature of system and very low output as only a tiny proportion of land area required is cultivated at any one time.

(b) To gain full marks candidates must comment on advantages and disadvantages of each change.

High yielding varieties (HYVs)
- Advantages – HYVs of staple food crops like rice and wheat have higher yields (fourfold in some areas) and grow more quickly so that more crops are harvested each year. This has meant a shift from subsistence farming towards more commercial farming with surpluses for sale.
- Disadvantages – HYVs are less drought-resistant, more susceptible to pests and disease and need large amounts of expensive fertilisers. Local people claim HYVs do not taste as good.

Mechanisation
- Advantages – The use of mini-tractors (rotavators) and small mechanised rice-harvesters instead of draught animals means farming is less labour-intensive, reducing labour costs and allowing amalgamation of uneconomic small fields and farms, and farming on a larger, more profitable scale.
- Disadvantages – richer farmers have benefited most. Poorer farmers have lost their land, causing unemployment and rural-urban migration.

Question 8: – Urban

(a) For Glasgow, candidates may refer to:
- Pedestrianisation and landscaping of CBD roads eg Buchanan Street, Argyle Street etc to reduce traffic flow in and around the CBD – to increase pedestrian safety and improve air quality and environment. Upgrading of CBD open space eg George Square.
- Diversification of city employment – much greater emphasis on tourist industry (significance of city-break holidays) leading to increased bed accommodation in new CBD hotels (Hilton, Radisson). Hotels can also tap into lucrative conference market given Glasgow's improved image as a tourist and cultural centre.
- Alteration of CBD road network – one-way streets (around George Square), bus lanes to discourage use of private transport and encourage use of public transport. Also achieved by increased metering and increased parking charges in and around CBD.
- Renovation and redevelopment of many CBD sites to provide modern hi-tech office space (Lloyds TSB, Direct Line etc) and residential apartments (Fusion Development, Robertson Street).
- Building of M8 and M74 extension all designed to keep traffic off CBD roads.
- Younger, more affluent population continues to be attracted to central city area by long-standing concentration of up-market pubs, clubs, cinemas etc (Cineworld in Renfrew Street).

(b) The description and explanation of urban landscape characteristics may include reference to environmental characteristics such as noise levels and pollution.

Inner City: Transition zone, first developed in the 19th century, mixture of old and newer buildings (recent regeneration to encourage more people to live there), high density of tenement or terraced housing, high-rise flats, derelict land and waste ground, redevelopment occurring, new houses of mixed type, lack of greenery and open space, environmental improvements exist in some areas etc.

Outer Suburbs: Modern late 20th/early 21st century semi-detached and detached housing, gardens and greenery around the areas, low density and high quality environment with well-planned street patterns of crescents and cul-de-sacs. Some outer-city council estates with flats, high rise and mixed housing. Growth of car ownership, commuting and demand for quieter and safer residential environments have led to the growth of these areas. Candidates may refer to land values.

GEOGRAPHY HIGHER
ENVIRONMENTAL INTERACTIONS
2012

Question 1 – Rural Land Resources

(a) Well annotated diagrams could be awarded full credit.Candidates should refer to surface and underground features for full marks, such as:
- limestone pavements
- sink/swallow/potholes
- dolines/shakeholes
- disappearing/resurgent streams
- dry valleys
- gorges
- scars and scree
- caverns
- stalagmites, stalactites, pillars.

Appropriate explanations should be provided for the formation of the features eg for a limestone pavement:
- limestone is a sedimentary rock laid down in layers (separated by bedding planes) under a tropical sea
- plate tectonics have raised the limestone while folding and faulting have resulted in joints (through pressure release)
- glacial erosion has scraped clear the overlying soil and exposed the limestone
- the joints/bedding planes are lines of weakness that can be enlarged due to the action of chemical weathering
- rainfall is a weak acid (carbonic acid) and the limestone (calcium carbonate) can therefore be dissolved by this rainwater
- this leads to deep gaps (grykes) and raised blocks (clints) in the exposed plateau or limestone pavement.

(b) Candidates must discuss at least two land uses to achieve full marks (not just tourism/recreation). Ideally candidates should identify the specific feature of the landscape and then go on to explain the opportunity it provides ie White Scar Caves and the guided tours, gift shop and café that are provided for tourists in the Ingleton area and the resulting jobs and money brought into the local economy.

Responses will vary according to the area chosen but opportunities might include:
Social – tourism, recreation, nature conservation.
Economic – farming, forestry, water supply, energy generation, quarrying.

(c) (i) Environmental conflicts in upland landscape usually include:
- air and noise pollution from traffic, quarrying…
- traffic congestion, heavy traffic damaging roads…
- erosion of footpaths, damage to fences and walls, litter…
- water pollution
- visual pollution from wind farms, ski-lifts/funicular railway, car parks, new buildings…
- impact on wildlife eg geese migration
- impact on tourism eg overhead Beauly to Denny power line

(ii) Precise points will obviously depend on the area chosen. To gain full marks candidates must comment on the effectiveness of their solutions/measures taken to resolve environmental conflicts.

Measures taken to resolve environmental conflicts might include:
- traffic restrictions in more favoured areas/at specific peak times eg one-way systems, bypasses or complete closures
- encouraging the use of public transport eg park and ride, minibuses, the use of alternative transport eg cycle paths and bridle ways
- separating local and tourist traffic, the use of permits (for access or parking) in some areas
- attempting to develop wider spread of 'honeypot' areas
- screening new buildings, car parks etc behind trees and only using local stone for buildings
- better visitor education
- burying power lines/development of off-shore energy production

Question 2 – Rural Land Degradation

(a) Candidate descriptions may include:
- high temperatures throughout the year – all stations have daily maximum temperatures above 30°C all year, rising to 44°C in Tombouctou in June
- isohyets show desert or semi-desert precipitation figures for much of northern Mali, with long dry periods from October to May
- even in south where precipitation is > 1000mm, there is little rainfall in winter
- as much of the country is desert or semi-desert, highly variable annual rainfall can also be assumed.

Candidates should explain why these climate patterns lead to the degradation of rural land. Points would include soil erosion from wind and infrequent and often heavy rainfall and impact of drought and desertification on vegetation.

(b) **North America** – most likely area Great Plains
For **Monoculture** – the growing of vast areas of wheat or cotton to supply worldwide demand stripped the land of the same nutrients year after year. This led to a decrease in fertility and moisture content and eventually a breakdown in the soil structure, leading to land degradation.
For **Farming marginal land** – ploughing of marginal land in the western plains during wet years left the soil fragile during dry years. Overgrazing of land by vast herds of cattle as demand for beef increased. Land became dry and compacted and vulnerable to wind and water erosion.

Africa north of the Equator – the Sahel
For **Deforestation** – rural-urban migration has led to a growing demand for wood for fuel.
Trees are already scarce in the area and further deforestation reduces interception and water storage leaving the soil dry and vulnerable to wind and water erosion.
For **Population increase** – desertification has caused people to migrate southwards within the Sahel leading to increasing population and increased pressure on already fragile land. Also many nomads have been forced to settle in villages, leading to greater food requirements and overuse of marginal land.

The Amazon Basin
For **Mining** – there are large deposits of gold, bauxite, iron ore, tin ore and diamonds in the Amazon Basin. In order to extract these minerals, large areas of the forest have been cleared. For instance, about one-sixth of Brazil's tropical rainforest (900,000 km²) has been cleared to mine the high quality iron ore found there. These mining activities have caused irreparable damage to large areas of land with gold mining methods poisoning soil and rivers. Tin miners rely heavily on hydraulic mining techniques, blasting away at river

banks with high-powered water cannons and clearing forests to expose potential tin deposits, leaving land totally degraded.
For **HEP** – The great rivers of the Amazon basin have a huge potential energy in the form of hydro-electric power. The Brazilian government had built 31 dams in the Amazon region by 2010. The amount of irreversible environmental damage they cause is huge. After the dam is built, the land slowly floods, driving the native Indians away from the river and eventually drowning their village and destroying the entire forest in the valley, endangering animal and plant species, sometimes making them extinct.

(c) **North America** – most likely TVA area or Dust Bowl.
For **Contour ploughing** – ploughing round, rather than up and down, slopes – rain has more time to infiltrate rather than form rills and gullies down slopes – the water soaks into the land providing extra moisture as well as preventing damage to the soil on the slope.
For **Shelter belts** – planting rows of trees at right angles to the direction of the prevailing wind – these act as a barrier for the land behind by reducing the force of the wind – the higher the barrier/trees the greater the protection.

Africa north of the Equator – the Sahel
For **animal fences** – movable fencing allow farmers to restrict grazing animals to specific areas of land and allow remaining land to recover. This allows farmers to move animals between fenced areas, reducing the dangers of overgrazing and trampling of soil and allowing soil and land to recover between grazing sessions.
For **"Magic Stones"** – This is a simple but very effective method of conserving soil. Diguettes are lines of stones laid along the contours of gently sloping farmland to catch rain water and reduce soil erosion. Diguettes allow the water to seep into the soil rather than run off the land. This prevents soil being washed away and can double the yield of crops such as groundnuts.

The Amazon Basin
For **Agroforestry schemes** – Agroforestry is the growing of both trees and agricultural/horticultural crops on the same piece of land. They are designed to provide tree and other crop products and at the same time protect and conserve the soil. It allows the production of diverse crops benefiting both land and peoples.
For **Purchase by conservation groups** – conservation groups, both national and international, aim to conserve soils by reforestation and the protection of existing forests eg the Amazon Region Protected areas (ARPAs) – created in 2002 by the Brazilian government in partnership with WWF, Brazilian Biodiversity Fund, German Development Bank, Global Environment Facility and World Bank – is a 10-year project aimed at increasing protection of the Amazon. By 2008, 32 million hectares of new parks and reserves were created in the Brazilian Amazon under ARPA, among them the 3.88 million-hectare Tumucumaque Mountains National Park, one of the world's largest national parks.

Question 3 – River Basin Management

(a) Description and explanation of need for water management might include:
  • Map Q3 indicates that the Zambezi River has many tributaries and the river basin has a very high drainage density leading to unpredictability of river flow which is dependent on when and how quickly snow melts in surrounding mountain areas.
  • Increasing population in Zambia and surrounding countries gives increasing demand for water for domestic, power, industrial needs.

  • Increasing demands from farmers for irrigation water to try and feed increasing population.
  • Rainfall figures for Mongu indicates seasonal nature of rainfall – extremely dry from May to September but high monthly figures for November to March – leading to flooding and also run-off of water that could be stored and used in dry months.
  • Temperature figures for Mongu indicate high temperatures throughout the year leading to very high evaporation rates. Monthly maximum temperatures peak at 34°C and never drop below 27°C.
  • There is a need to regulate flow of river to prevent flooding during peak discharge and to keep water level high enough for navigation in dry months.
  • There are 8 countries within the Zambezi River basin and they will have conflicting demands for a share of the river's water.

(b) Physical factors might include:
  • Geologically stable area away from earthquake zones/fault lines.
  • Solid rock foundation for weight of dam.
  • Narrow valley cross-section to reduce dam length.
  • Large, deep valley to flood behind dam to maximise amount of water storage.
  • Lack of permeability in rock below and around reservoir to prevent seepage.
  • Low evaporation rates.
  • Large catchment area above dam to provide reliable water supply.

Human factors might include:
  • Cost of dam construction.
  • Proximity of urban area for water and electricity.
  • Proximity of agricultural areas for irrigation.
  • Cost of displacing people.
  • Cost of compensating farmers and home owners.
  • Impact on communications.
  • Unstable political situation and internal conflict eg Zimbabwe.

(c) Candidates must refer to all 6 sections for full marks. Answers will depend on the river basin chosen. However, for the Colorado River they might include:

Social benefits:
  • Fresh water supply for growing desert cities eg Phoenix.
  • Better standard of living in hot, dry climate with air conditioning, swimming pools, landscaping etc.
  • Areas at reservoirs, eg Lake Mead, give opportunities for tourism, water sports, fishing etc.
  • Regulation of river greatly improves flood control on river.

Social adverse consequences:
  • People had to be moved off their land as valley areas were flooded.
  • Loss of burial sites and other Native American sacred areas.
  • Disagreements between states and countries with regard to allocation of water from river.

Economic benefits:
  • Cheap HEP attracted industries eg electronics to take advantage of the area's cheap land and low taxes.
  • Benefited tourist industry with reliable water supply – attractions like the Grand Canyon, gambling in Las Vegas, Hoover Dam etc.
  • Expansion of irrigated land led to agribusiness-style farming.

Page 184 — OFFICIAL SQA ANSWERS TO HIGHER GEOGRAPHY

Economic adverse consequences:
- Huge cost of building the dams eg Central Arizona Project cost $6 billion.
- High cost of maintaining dams, power plants and irrigation channels.
- Subsidised water for farmers has led to water wastage and the growing of crops that could be produced cheaper elsewhere.

Environmental benefits:
- Reservoirs provide sanctuaries for waterfowl and wading birds like the blue heron.
- The National Recreation Area around Lake Mead has more than 250 species of birds.
- Reliable seasonal water flow for plants and animal life.

Environmental adverse consequences:
- Water in river and on farmland becomes saline with high evaporation rates – farmers downstream have to switch to more salt-tolerant crops.
- Change in river regime has caused the loss of many animal habitats eg the drying up of Colorado delta area where there used to be a great variety of birdlife.
- Huge amount of water loss by seepage through the sandstone rocks around Lake Powell. Scenic attractions like the Rainbow Bridge are being affected by the high water levels in Lake Powell.

Question 4 – Urban Change and its Management

(a) Answers will depend upon the Developed Country that is chosen, but for Australia answers might include:
- Concentration of major cities in south-east Australia between Melbourne and Sydney due to early penal colonies and mineral deposits eg gold, zinc and bauxite.
- Coastal cities bordered by eastern mountain ranges like Blue Mountains act as barriers to extension.
- The interior of Australia has no large cities due to desert.
- Perth on the coast of Western Australia, due to natural resources and agricultural centre for the state.
- Darwin is a city in the more tropical north, an early trading port.
- The capital city, Canberra, is inland, planned to attract investment inland.
- Hobart is on the island of Tasmania to the south-east of the mainland.

(b) Answer will depend on the city chosen, but for Melbourne the following might be suggested:
- Natural, sheltered bay ie Port Philip Bay suitable for port development.
- Located at estuary of Yarra River, again suitable for shipping and dock development.
- Developed as State capital of Victoria.
- Nearby beaches for water based leisure and tourism.
- On a coastal plain with flat land to allow urban growth.
- South of mountain range ie Australian Alps.
- Route centre along coastal plains and to the north through Kilmore Gap.

(c) (i) Problems caused by urban sprawl might include:
- Using up valuable recreational land and farmland.
- Threatening wildlife habitats and removing green areas and open land.
- Increased commuting leading to traffic congestion, more road building and increased levels of air pollution.
- Longer travelling times from suburbs to CBD.
- Building and services in inner city urban areas not being used or becoming run down or derelict eg housing, schools, factories and shopping centres.
- Flooding issues caused by development on the floodplain.

(ii) Problems caused by urban sprawl might be resolved in the following ways:
- Protecting the Green Belt through legislation and planning controls.
- Prohibit urban development in Green belt areas.
- Designate urban growth corridors.
- Redevelop inner urban areas with modern affordable housing, high quality shopping areas and create hubs for entertainment.
- Encourage industry onto brownfield sites through grants and incentive schemes.
- Improve public transport systems reducing need for additional motorways and related infrastructure eg park and ride systems, tramways.

(d) (i) Answers may include the following points:

Rural 'push' factors:
- Low income from farming and related work.
- Lack of employment in manufacturing and service industry.
- Lack of education and low literacy levels.
- Poor health facilities.
- Higher levels of disease and malnutrition.
- Poor sanitation and lack of utilities eg electricity, water, gas.
- Poor quality of infrastructure.
- Civil unrest.
- Rural land degradation.
- Environmental disasters eg drought, earthquakes.

Urban 'pull' factors:
- Mainly reversals of above, but may also include entertainment and bright lights.
- Informal employment opportunities, improved housing, development of social communities.

Generic reasons can also be accepted ie population growth with higher birth rates and falling death rates, immigration from neighbouring countries and movement into cities by refugees.

(ii) Answers may include:
- Impoverished and overcrowded areas which may lack public utilities and amenities of water supply, electricity and sewerage.
- Erratic power supply.
- Semi-urban peripheral areas with poor housing quality and poor economic opportunities.
- Squatter settlements located on steep slopes and lagoons.
- Areas lacking basic services eg schools, piped water and hospitals.
- High incidence of disease.
- High rates of unemployment, growth of black market and a 'grey' economy, crime, drugs and prostitution increasing.
- Problems of waste management and disposal.
- Water pollution from toxic industrial waste and sewage effluent.
- Unhealthy city dumps and landfill sites.
- Air pollution caused by chronic traffic congestion and uncontrolled industrial emissions.

Question 5 – European Regional Inequalities

(a) (i) Credit should be awarded for candidates noting that Objective 1 status was awarded to Europe's peripheral areas, eg NW Ireland, SW England, Wales, Portugal and Spain, S Italy, Greece, N Sweden, N Finland, the eastern part of Germany and the 2004 Eastern bloc EU members.

Convergence zones are very similar adding Bulgaria and Romania (the most recent EU members) and no longer providing this type of support to Ireland, areas of Northern Spain, Sardinia, Sweden and Finland.

(ii) Countries would benefit from Convergence Region funding though the following:
- Support for infrastructure improvements.
- Support for employment training and education.
- Support for production/manufacturing sectors.
- Environmental protection.
- Improving access to the peripheral areas.
- Improving IT, literacy and numeracy.

Other European measures which candidates may refer to include:
- European Regional Development Fund (ERDF), which provides a wide range of direct and indirect assistance to encourage firms to move to disadvantaged areas, eg loans, grants, infrastructure improvements.
- European Investment Bank (EIB) provide loans for businesses setting up in disadvantaged areas.
- European Social Fund (ESF) assists with job retraining and relocation.

(b) (i) Description might include the marked differences in the four indicators across the 27 member states. Figures should be quoted and countries named. Bulgaria, Romania, Latvia and Poland stand out as being particularly disadvantaged compared to countries such as Luxembourg, Ireland, Netherlands and Sweden. Candidates may wish to describe the pattern with reference to year of membership and may note that some indicators such as employment rates do not show such a strong difference between richer and poorer countries. Italy has a significant lower employment rate (58%) compared to the other original EU members. Candidates may also note other exceptions to the general pattern eg Slovenia's higher HDI ranking and GDP per capita, in comparison to other newer members of the European Union.

(ii) Candidates should refer to both physical and human factors to account for the variation in level of development, including factors such as:

Physical factors
- Relief/geology
- Climate/water/resources
- Soil quality/soil erosion
- Natural Disasters

Human Factors
- Remoteness/isolation/communications
- Employment opportunities
- Historical access to raw materials
- Development of industry/technology
- Land tenure
- Agricultural development
- Population density/location and proximity to markets
- Unskilled labour, poorly educated workforce
- Some Eastern countries later in joining due to fall of communism so not gaining benefits until later

Candidates may refer to specific examples eg the newer Eastern periphery countries whose reasons for lack of prosperity might include:
- Less favoured area climatically – very cold continental winters
- More remote from major political centres of Europe – London, Strasbourg, Paris, Brussels.

- Less well served by transportation links – away from major motorway networks, away from 'hub' airports like Schiphol, London and Paris airports and away from major container ports like Europort with worldwide trade links.
- Legacy in some cases of high unemployment or underemployment and high percentage with relatively poor living standards and loss of younger workforce through migration to wealthier EU countries.

(c) For maximum marks some comment must be made on the effectiveness of the strategies named. Answers will depend on the country chosen.

National Government Measures include:
- Regional development status, Enterprise Zone status, capital allowances, training grants, assistance with labour costs.
- Specific assistance to former coal mining/iron and steel areas
- Intervention of national government resulting in the relocation of major government employers or state owned firms to disadvantaged areas eg Fiat to Southern Italy, DVLA in Swansea, MoD in Glasgow.
- In Italy the Cassa per il Mezzogiorno would be a key policy.

Question 6 – Development and Health

(a) Economic indicators could include:
- Gross Domestic Product per capita in US $.
- Average Annual Income per capita in US $ or GB £.
- Percentage of working population employed in, say, the Primary sector.

Social indicators could include:
- Adult literacy rates (%).
- Average life expectancy at birth in years.
- Number of cars/TV sets/telephones etc per 1000 people.

Candidates should explain what the indicator demonstrates about the level of development in a country eg GDP per capita is a measure of the value of goods and services and demonstrates the development of industry etc.

(b) Answers may refer to the likes of:
- Oil-rich countries such as Saudi Arabia, Brunei, the UAE or to relatively well-off countries like Malaysia which are able to export primary products such as tropical hardwoods, rubber, palm oil and tin as opposed to poorer nations such as Burkina Faso or Chad which lack significant resources.
- Newly Industrialising Countries (NICs) eg China, South Korea, Taiwan are able to earn substantial amounts from steel-making, shipbuilding, car manufacturing, electrical goods, toys, clothing etc. They have been able to benefit from their population's entrepreneurial skills and low labour costs.
- Some countries such as Brazil and India have both resources and growing manufacturing industries.
- The expansion of tourism has helped to improve living standards/create new job opportunities in countries like Thailand, Jamaica, Kenya, Sri Lanka and earns valuable foreign currency.
- Many countries are afflicted by recurring natural disasters which restrict development/hamper progress eg drought in sub Saharan Africa (Mali, Chad, Burkina Faso...) – floods/cyclones in Pakistan/Bangladesh – hurricanes in the Caribbean – tsunamis in Sri Lanka, Indonesia.
- Political instability – eg recent disruptive civil wars in places such as Sudan/Rwanda/Somalia/Liberia/Sierra Leone or larger-scale conflicts in Iraq or Afghanistan have also had a negative impact. Widespread corruption and

mismanagement have accounted for the marked decline of Zimbabwe's economy and are a continuing problem in many other African nations.

(c) There may be some overlap between human and physical factors.

(i) For Malaria – Physical factors:
  - Female Anopheles mosquito.
  - Hot and wet climates such as those experienced in the Tropical Rainforest or Monsoon areas of the world.
  - Temperatures between 15°C and 40°C.
  - Areas of shade in which the mosquito can digest blood.
  - Stagnant water.

Human factors:
  - Nearby settlements to provide a 'blood reservoir'.
  - Suitable breeding habitat for the mosquito – areas of stagnant water such as irrigation channels, reservoirs or poor drainage that leaves standing water uncovered eg tank wells, Irrigation channels, water barrels, padi fields.
  - Exposure of bare skin.
  - Increased trade and tourism.
  - Not completing courses of drugs.

(ii) Strategies used to combat the spread of Malaria may include:
  Trying to eradicate the mosquito:
  - Insecticides eg DDT and now Malathion.
  - Mustard seeds thrown on the water that becomes wet and sticky so dragging the mosquito larvae under, drowning them.
  - Egg-white sprayed on the water creates a film which suffocates the larvae by clogging up their breathing tubes.
  - BTI bacteria grown in coconuts – the fermented coconuts are broken open after a few days and thrown into the mosquito-larvae infested ponds – the larvae eat the bacteria and have their stomach linings destroyed.
  - Larvae eating fish introduced to ponds.
  - Draining swamps, planting eucalyptus trees that soak up excess moisture, covering standing water.
  - Genetic engineering eg of sterile males.

Treating those suffering from malaria:
  - Drugs like chloroquin, larium and malarone.
  - Qinghaosu extracted from artemisinin plant – a traditional Chinese cure
  - Continued search for a vaccine – not available as yet.

Education programmes in:
  - The use of insect repellents eg Autan.
  - Covering the skin at dusk when the mosquitoes are most active.
  - Sleeping under an insecticide treated mosquito net.
  - Mesh coverings over windows/door openings.
  - WHO 'Roll Back Malaria' campaign.
  - The Bill and Melinda Gates Foundation.

(d) Examples of Primary Health Care (PHC) strategies may include:
  - Use of barefoot doctors – trusted local people who can carry out treatment for more common illnesses – sometimes using cheaper traditional remedies.
  - Use of ORT (Oral Rehydration Therapy) to tackle dehydration – especially amongst babies. This is an easy, cheap and effective remedy for diarrhoea/dehydration.
  - Provision of vaccination programmes against disease such as polio, measles, cholera. Candidates may also refer to PHC as based on generally preventative medicine rather than (more expensive) curative medicine.
  - The development of health education schemes in schools, community plays/songs concerning AIDS, with groups of expectant mothers or women in relation to diet and hygiene. Oral education being much more effective in illiterate societies.
  - Sometimes these initiatives are backed up by the building of small local health centres staffed by visiting doctors.
  - PHC can also involve the building of small scale clean water supplies and Blair toilets/pit latrines – often with community participation.
  - The use of local labour and building materials is often cheaper; it also provides training/transferable skills for the participants and gains faster acceptance/usage in the local and wider community.

GEOGRAPHY HIGHER
PHYSICAL AND HUMAN ENVIRONMENTS
2013

SECTION A

Question 1 – Lithosphere

(a) **Surface feature – Limestone Pavement**
- areas of bare limestone scraped clear of soil and glacial drift
- limestone surface exposed to chemical weathering
- joints formed in limestone as it dried out or as pressure was released
- lines of weakness prone to chemical weathering called carbonation. The limestone is dissolved by rainwater (weak carbonic acid)
- deep gaps (grykes) and blocks (clints) make up the distinctive landscape from the original horizontal bedding plane
- credit will be given to solution and biological activity

Underground Features – Stalagmites and stalactites
- found in cavern systems where underground water is rich in lime
- water percolates through the joints and bedding planes as the rock is permeable
- in a cave roof icicles of calcite form from dripping water which evaporates depositing crystalline lime ie stalactites
- lime is deposited on the floor and is more rounded ie stalagmites
- features are called dripstone deposits
- pillars form when stalactites and stalagmites meet

(b) Conditions and process which encourage the formation of scree slopes will include:
- steep and bare rock faces
- well jointed rock face with lines of weakness
- cold climate where temperatures often fall below freezing at night
- freeze-thaw action or frost shattering
- water collects in cracks and freezes and expands by 9% exerting pressure on rock
- repeated freeze-thaw action splits rock into large sharp fragments
- fragments break off and move downhill by gravity
- accumulate at the base of a cliff as scree or talus slope
- grading of scree material upwards from larger to smaller due to weathering

Question 2 – Atmosphere

(a) Candidates should explain both the term reflection (from the Earth's atmosphere and surface) and absorption within the Earth's atmosphere.

Reflection and scattering reduces the amount of solar energy by about 30%. This is known as the albedo effect. Approximately 20% is reflected by clouds, 5% is scattered by gas particles in the air and 5% is reflected from the Earth's surface. Reflection therefore varies dependent on cloud cover and also the covering at the Earth's surface as darker forest surfaces absorb more radiation than snow and ice surfaces which reflect more of the incoming radiation. The spatial variation is emphasised by equatorial forests and polar ice caps.

Absorption by the atmosphere reduces the solar energy by (20%), through clouds (3%) and by dust, water vapour and other gases (approx. 17%).

(b)
- Carbon dioxide: from burning fossil fuels – road transport, power stations, heating systems and from deforestation (particularly in the rainforests) and peat bog reclamation/development (particularly in Ireland and Scotland for wind farms).
- CFCs/PFCs: from aerosols, air-conditioning systems, refrigerators, polystyrene packaging, production of aluminium, etc.
- Methane: from rice padis, landfill sites, (almost half of UK's methane emissions), animal dung, oil exploration, permafrost melting in tundra areas and belching cows.
- Nitrous oxides: from vehicle exhausts and power stations.
- Sulphur Hexafluoride: from electrical substations, magnesium smelters.
- Hydrofluorocarbons (HFCs): potent greenhouse gases used to replace CFCs in refrigeration, air conditioning and the production of insulating foams.
- Global 'dimming' from sulphate aerosol particles, atomic bomb detonations and aircraft contrails. Polluted clouds are more reflective/absorbent than unpolluted clouds, increasing reflection/ absorption in the atmosphere and therefore cooling.

NB there were 6 man-made greenhouse gases included in the Kyoto protocol (Carbon Dioxide, Methane, Nitrous Oxide, Hydro fluorocarbons, PFCs and Sulphur Hexafluoride). Many of these are more powerful as greenhouse gases than CO2.

Question 3 – Urban

a) The following characteristics may be noted:

Area A
- Grid iron street patterns
- Terraced housing
- Older (19th century) development, next to CBD, inner city
- Lack of open space
- Main roads nearby – B3238
- More churches – 496552, 494554
- Railway lines nearby near to industrial area at port – 493537

Area B
- Cul-de-sacs, curvilinear street patterns with crescents
- Modern (20th century) development, on edge of city
- Commuter/dormitory settlement
- Detached and semi-detached housing with gardens
- Open space
- Minor roads only through residential area
- access to A386 for commuters
- Woodland on southern edge could be credited for recreational use
- easy access to paths to countryside
- long distance footpath nearby for walking – West Devon Way

Reasons relate to different periods of development, which in turn relate to location. Area B is modern (probably 20th century) suburban development, commuters dependent on road transport and being on edge of city, lower housing density is possible on cheaper land.

Area A is inner city area which grew in 19th century adjacent to city centre and industrial/docklands area to south, in pre-car era hence high housing density.

(b) The likely impact of a new out of town regional shopping centre could include references to such points as:
- loss of custom for shops
- loss of customers for other services such as cinemas and restaurants

- consequent closure of shops/services and dereliction/empty properties
- relocation of shops and services to the new site
- probable revitalisation of traditional shopping streets in the city centre in order to compete/'keep up'
- changes to local planning controls, initiatives to bring back customers to the CBD.
- loss of council revenue from services e.g. car parks.

Question 4 – Rural

(i) **Intensive Peasant Farming**
- Traditional high labour input although this is beginning to decline as poorer farmers are forced off the land
- Small capital input although this is increasing with amalgamation of uneconomic holdings and increased use of machinery
- Small parcels of land but this is also increasing with amalgamation
- Large output due to intensive nature of system with maximum use of land available.

Commercial Arable Farming
- Labour force small and declining with increased use of large machines as agribusiness takes over from family farms
- High input of capital, used for machinery, irrigation, pesticides, fertilisers and infrastructure
- Very large areas of land required for effective operation of large farm machinery
- Large output is related to huge area involved rather than particularly high output per hectare.

(ii) Descriptions might include:

Intensive Peasant Farming
- improved irrigation
- increased farm sizes and larger fields – amalgamation of small uneconomic holdings and consolidation of fragmented fields as a result of land reform
- greater use of modern pesticides and fertilisers
- increased mechanisation – the use of mini-tractors (rotovators) and small mechanised rice-harvesters instead of draught animals
- The widespread adoption of higher yielding/faster maturing new varieties of rice – the impact of the 'Green Revolution'
- 'green revolution' type changes eg development of hybrid seeds
- use of appropriate technology
- increasing export of farming produce
- the formation of farming co-operatives.

Commercial Arable
- amalgamation of farm holdings as family farms are taken over by agribusiness
- part time farming and co-operatives have increased
- greater use of contractors for harvesting
- diversification of crops away from wheat to eg sunflowers as markets change
- increase in organic farming
- increased use of more carefully managed irrigation schemes
- increased awareness of soil conservation methods

The impact of these changes might include:

Intensive Peasant Farming
- greater amount of food has reduced malnutrition and starvation
- surplus crop may be sold, improving quality of life
- increased mechanisation may lead to reduction in farm labour
- migration of farm workers to urban areas and impact on

demography of rural areas
- consolidation of farms may also lead to larger fields, increased mechanisation and drift to cities
- improved infrastructure including increased electrification and better roads improving access to markets
- co-operatives have provided farmers with several benefits, easier access to machinery, cheaper credit facilities, bulk purchasing of inputs and improved marketing opportunities
- a shift from subsistence farming towards more commercial farming with small surpluses for sale
- increased use of insecticides, pesticides and fertilisers may impact on the environment and humans.

Commercial Arable Farming
- decline in rural population as family farms are taken over
- abandoned homesteads and decline in rural services such as schools
- young families tend to move out so the population becomes an ageing one
- wider range of crops means the endless expanse of crops is a less common sight on the Great Plains
- the diversification of cropping has helped secure income as have cooperatives (economies of scale from cooperatives)
- planting of eg sunflowers as part of strip cropping has decreased soil erosion
- increase in organic farming has meant eg less algae in some local rivers.

SECTION B

Question 5 – Hydrosphere

(a) Description could include:
- Lower course section of river
- Narrow valley 5261
- Meanders eg 519607
- Tributaries/confluences, (eg River 519567)
- Braiding/islands (eg 522587), pond (NOT ox-bow lake) at 520586
- River cliff (524597)
- Widening valley – flat floodplain 518584
- References to the height of the land, steepness of the valley sides
- Direction of flow southerly
- Credit speed of river if linked to map evidence
- Tidal limit at 518571
- Mudflats at mouth, 514558

(b) Points could include:
- Development of pools and riffles (differences in speed and depth)
- Erosion on the outside (concave bank) of bends due to faster flow forms river cliffs
- Helicoidal flow removing material
- Deposition on the inside (convex bank) due to slower flow, formation of point bars/river beaches
- Migration of meanders downstream

Question 6 – Biosphere

(a) For full marks, description and explanation needed.

A Strandline: Likely plant types include Sea Sandwort, Sea Rocket, Saltwort. Salt tolerant species, able to withstand dessicating effects of the sand and wind.
Able to cope with periodic immersion in sea water. Have also to adapt to alkaline conditions (high pH) because of high concentrations of shell fragments along the shore.

Embryo Dunes: Examples of plants such as Sand Couch, Lyme Grass, Sea Rocket. These pioneer species grow side / lateral roots and underground stems called rhizomes which bind the sand together. These grassy plants can also tolerate occasional immersion in sea water.

B Fore Dunes: Examples of plants such as Sea Holly, Sea Bindweed, Sand Sedge.
Slightly higher humus content from decayed plants and lower salt content (further from the sea) allows these species to further stabilise the dune and allow the establishment of Marram Grass which becomes a key plant in the build up of the dune.

Main ridge (Yellow Dunes): mainly Marram Grass and Sea Lyme Grass but also Sand Fescue, Sand Sedge, Sea Bindweed and Ragwort. Both humus content and soil acidity have increased at this point. Being xerophytic, Marram Grass thrives on the drier mobile sands and becomes the dominant species, aligning itself with the prevailing wind to prevent moisture loss / reduce transpiration and can keep pace with being buried under deposits of sand thanks to its long, creeping rhizomes spreading laterally and vertically.

C Grey Dunes: Plants often present include: Bird's Foot Trefoil, lichens, mosses, heather, Sea Buckthorn. Marram Grass dies back, contributing humus. As a result of leaching and the build up of humus, the soil is considerably more acidic and damper, allowing a wider range of plant species to grow in this more sheltered location.

Dune slacks: Reeds and Rushes, Cotton Grass, small Willows and Alders. The damp, low-lying hollows have a much higher water-table, especially in winter and support a hydrophytic (water tolerant) vegetation cover. Increased organic matter, shelter from the dune ridge in front and, being further inland, a less saline environment also contribute to a wider range of plants. As a result of water logging and the build up of humus, the soil is also more acidic.

(b) Annotations could include:
- Dark brown A horizon with thick black layer of mull humus and loamy texture
- Abundant leaf litter from deciduous woodland
- Lighter brown in upper B horizon with some staining from organic matter above
- Dark brown/red brown in lower B horizon
- Presence of soil biota eg earthworms
- Well drained but an iron pan may be present in the B horizon caused by moderate leaching
- Indistinct horizon boundaries – presence of organisms vertically mixing soil
- Clay particles in C horizon above the bedrock

SECTION C

Question 7 – Population

(a) Description could include:
- high birth rate, shown by the broad base and high percentage in the younger age groups
- high death rate, shown by the rapidly narrowing pyramid
- low life expectancy and high infant mortality rate
- relatively small economically active population between 15 and 65 age groups

Explanations could include:
- high birth rate due to lack of family planning, contraceptives and sex education
- HBR also due to a need for large families to help in farming and also due to the high infant mortality rate, lack of education for girls
- High death rate will be due to high infant mortality, occurrence of 'killer' diseases like malaria
- High levels of HIV/Aids
- HDR due to malnutrition causing susceptibility to disease
- HDR due to poor medical availability
- Lack of clean water and sanitation

(b) Consequences on the economy might include:
- positives from an increased economically active population providing a healthy work force
- the more likely negatives will include increasing population leading to increased unemployment and a greater need to borrow money and increase National Debt
- a greater reliance on overseas aid
- less surplus produce available to sell abroad

Consequences on the welfare of the citizens:
- positives from population living longer and being healthier eg fitter and more effective workforce
- rural migration to towns to find work
- pressure on larger towns to accommodate rising population which may lead to growth of shanty towns
- rising population will lead to pressure on services like sanitation, schools and hospitals
- rising population may lead to pressure on food supplies leading to increased malnutrition and disease
- lack of work may lead to increased emigration to neighbouring countries or overseas
- rising population may put pressure on all resources eg. fuel, building materials etc.

Question 8 – Industrial Geography

(i) Main characteristics that might be included in a description/explanation of a 'new' industrial landscape.
- Lower, smaller, modern buildings – mostly single storey and often with large windows to allow in plenty of light.
- Buildings are well planned/spaced out with trees and grassy areas and even ornamental lakes/ponds included in the layout to provide a more attractive working environment and create a favourable image to prospective investors/clients.
- Usually located on purpose-built industrial estates or Science/Business Parks commonly on Greenfield sites on the edge of towns/cities where land is relatively cheap and there is room for car parking and for future expansion.
- Usually close to major roads such as dual carriageways or motorways for ease of transport of the finished products to markets /ports, for bringing in raw materials/component/ sub-assemblies and for the convenience of to-day's more mobile, car-owning workforce.

- Similar sorts of industries/firms in similar looking buildings often locate on the same site to benefit from an exchange of ideas and information. Many of these businesses are connected with information, high technology and electronics industries and will have direct links with universities (often situated close by) for research and development purposes and to remain successful and competitive.

(ii) Answers will vary and will be determined by the industrial concentration chosen. For South Wales, steps taken to attract new industries and inward investment might include:

National Government
- creation of Enterprise Zones (Swansea, Milford Haven) and their associated benefits
- designation of Development Area status for old coal mining areas
- setting up of Welsh Development Agency (WDA) in 1976, to attract high quality investment into Wales
- Urban Development Corporation (UDC) in Cardiff and its associated benefits
- Improved infrastructure – the Heads of the Valleys Road
- construction of New Town, Cwmbran
- relocation of specific government offices, eg DVLA in Swansea
- encouraging inward investment from abroad, eg Sony, Bosch, Lucky Goldstar

European Union
- E.U. (creation of EU itself provides huge European market for goods)
- joining EU opens up a huge source of funds available to outlying areas – ERDF (European Regional Development Fund), EIB (European Investment Bank), ESF (European Social Fund) etc and their associated benefits
- Cohesion Fund – aimed at states whose Gross National Income (GNI) is <90% of EU average

GEOGRAPHY HIGHER ENVIRONMENTAL INTERACTIONS 2012

Question 1 – Rural Land Resources

(a) For an answer to achieve full marks, well annotated diagrams must be used. For a corrie points could include:
- snow accumulates in north/east facing hollow due to lack of melting
- successive layers of snow compress into ice/neve
- ice moves downhill under gravity
- freeze-thaw weathering occurs on the backwall
- rocks embedded in ice grind away at bottom of the corrie
- abrasion carves out armchair-shaped depression due to rotational movement
- rate of erosion decreases at edge of corrie leaving a rock lip

(b) Answers are expected to link these opportunities to the physical landscape and answers must mention both social and economic opportunities for full marks.

Explanation can be developed from:
Social Opportunities
- mountaineering, hill walking and skiing
- forest walks, picnic sites and orienteering
- sailing, fishing and other water sports
- nature conservation

Economic opportunities
- tourism and associated employment and profits
- development of hotels, bunkhouses and campsites
- hill sheep farming
- forestry plantations
- HEP and water supply
- quarrying

(c) (i) Answers should be able to describe/explain the environmental conflicts including:
- traffic congestion on narrow country roads, in honeypot sites, car parks etc
- air, noise and water pollution (eg from traffic, some water sports, quarrying)
- footpath erosion, damage to walls, fences etc and other forms of visual pollution (eg unsightly visitor centres, the Cairngorm funicular railway, caravan/campsites)
- some visitors may cause problems for farmers and landowners (eg litter, animal disturbance)

(ii) To achieve full marks candidates must refer to the effectiveness of their solution. Solutions might include a variety of environmental conflicts depending on the area chosen but for traffic congestion it could include:
- Traffic restrictions in more favoured areas/at specific peak times eg one-way streets, bypasses, the use of permits or complete closures eg the Goyt Valley Traffic Scheme or separate 'tourist routes'
- Encourage the use of public transport eg park and ride, minibuses and the use of alternative transport eg cycle paths and bridle ways

Question 2 – Rural Land Degradation

(a) The three main processes can be described from the reference diagram:
- Surface Creep – the slow movement of the larger particles across the land surface.
- Saltation – the 'bouncing' along of lighter particles.
- Suspension – the lightest particles (dust) being blown in the air.

- The explanation should focus on the principle that the wind can move smaller (lighter) particles more easily than larger (heavier) particles – hence the difference in process.
- The largest (and heaviest) particles (stones and boulders) will not be moved by the wind.

(b) For the Dust Bowl, answers may include:
- Use of techniques better suited to the moister eastern states.
- Monoculture, especially of wheat or demanding crops (cotton) depleted the soil of moisture and nutrients.
- Deep ploughing of fragile soils (previously these had been held in place by natural grasslands).
- Marginal land ploughed – particularly in wet years – leaving them in a fragile condition in dry years.
- Ploughing downslope creating opportunities for rill erosion.
- Farm sizes being too small so forcing farmers to over crop – particularly when prices were low and therefore income was low.
- Overuse of irrigation leading to salination of soil in places.

For the Tennessee Valley:
- Removal of shelter belts leading to increased risk of wind erosion.
- Mining and farming cleared the natural vegetation and led to soil erosion.
- Overcropping had already weakened the soil.
- Lack of fertiliser caused the soil to lose its structure and become vulnerable to erosion.

(c) For full marks, effects on the people, economy and the environment must be included.

For Africa, north of the Equator, answers for the impact on people and the economy may include:
- Crop failures and death of livestock, reducing food supply, leading to malnutrition and famine.
- Increased infant mortality rates / death rates eg Ethiopia, Sudan
- Collapse of the traditional nomadic way of life
- Large scale rural migration into overcrowded urban areas, causing more pressure and the growth of shanty towns
- Conflict within and sometimes, between countries as people move and re-settle –growth of large refugee camps
- Countries increasingly reliant on international aid

Environmental effects may include:
- Soil structure deteriorates due to over-cropping and over-grazing
- Wind erosion can remove large amounts of dried out soil
- Advance of the Sahara desert – desertification
- Water tables lowered
- Torrential rains can lead to gully erosion
- Intensified drought due to the albedo effect

For the Amazon Basin, answers for the impact on people and the economy may include:
- Destruction of the way of life of the indigenous people eg clashes between the Yanomami and incomers
- Destruction of the formerly sustainable development eg rubber tappers and Brazil Nut collectors
- Clashes between competing groups eg the violent death of Chico Mendez allegedly at the behest of ranchers
- Creation of reservations for indigenous people
- Increase in 'western' diseases and alcoholism

Environmental effects may include:
- Adverse effect on the nutrient cycle in the rainforest
- Leaching of minerals, removal of top soil and increased laterisation
- Increased surface run-off, flooding and silting up of rivers

- Loss of biodiversity with danger of extinction in some cases
- Loss of potential life-saving drugs
- Increased risk of climate change

(d) (i) Answers should provide reasonably detailed information about farming methods and include some explanations eg: Shelter belts on low-lying land affected by strong winds are rows of trees grown across the direction of the prevailing wind. They act as a barrier to slow down winds and protect the soil.

For North America, answers may include:
- Crop rotation
- Diversification of farming types
- Keeping land under grass or fallow
- Trash farming / stubble mulching
- Replanting shelter belts
- Strip cultivation and intercropping
- Contour ploughing
- Terracing
- Use of natural fertilisers
- Soil banks
- Improved irrigation

Effectiveness may include eg – most likely TVA area or Dust Bowl.

For **Contour ploughing** – ploughing round, rather than up and down, slopes – rain has more time to infiltrate rather than form rills and gullies down slopes – the water soaks into the land providing extra moisture as well as preventing damage to the soil on the slope.

For **Shelter belts** – planting rows of trees at right angles to the direction of the prevailing wind – these act as a barrier for the land behind by reducing the force of the wind – the higher the barrier / trees the greater the protection.

For **Africa, north of the Equator**, answers may include:
- Diguettes or 'magic stones'
- Dams built in gullies
- Animal fences
- Dune stabilisation

(ii) Effectiveness may include eg

For **Animal fences** – movable fencing allows farmers to restrict grazing animals to specific areas of land and allows remaining land to recover. This allows farmers to move animals between fenced areas, reducing the dangers of overgrazing and trampling of soil and allowing the soil and land to recover between grazing sessions.

For **"Magic Stones"** – This is a simple but very effective method of conserving soil. Diguettes are lines of stones laid along contours of gently sloping farmland to catch rain water and reduce soil erosion. Diguettes allow the water to seep into the soil rather than run off the land. This prevents soil being washed away and can double the yield of crops such as groundnuts.

(i) For the **Amazon Basin**, answers may include:
- Agro-forestry schemes
- Crop rotation
- Purchase by conservation groups
- Return land to traditional farming

(ii) Effectiveness may include eg

For **Agroforestry schemes** – Agroforestry is the growing of both trees and agricultural / horticultural crops on the same piece of land. They are designed to provide tree and other crop products and at the same time protect and conserve the soil. It allows the production of diverse crops benefiting both land and peoples.

For **Purchase by conservation groups** – conservation groups, both national and international, aim to conserve soils by reforestation and the protection of existing forests eg the Amazon Region Protected areas (ARPAs) – created in 2002 by the Brazilian government in partnership with WWF, Brazilian Biodiversity Fund, German Development Bank, Global Environment Facility and World Bank – is a 10 year project aimed at increasing protection of the Amazon. By 2008, 32 million hectares of new parks and reserves were created in the Brazilian Amazon under ARPA, among them the 3.88 million-hectare Tumucumaque Mountains National Park, one of the world's largest national parks.

Question 3 – River Basin Management

(a) Description should include reference to the general patterns / numbers of rivers, and should refer to the directions of flow. Explanation should refer to the fact that drainage basins are determined by the location of the main continental watersheds and that major rivers rise in the main mountain ranges that have greater precipitation, eg the Rockies and Appalachians in North America.

Description and explanation for North America river basins might include:
- west-flowing rivers are fed from the western side of the continental divide. Rivers like the Columbia-Snake and the Colorado flow west into the Pacific Ocean
- north-flowing rivers drain to the Arctic Ocean or to Hudson Bay and are fed from the Canadian Shield
- the St Lawrence system is fed from the Great Lakes areas and flows east to the Atlantic Ocean
- most of south-eastern USA is dominated by the Mississippi and its tributaries which are fed from the Rockies in the west and the Appalachians in the east and flow to the Gulf of Mexico

(b) Descriptions and explanation of need for water management might include:
- Map Q3D indicates that the Mississippi River has many tributaries (such as the Missouri, Platte and Tennessee named on map Q3A), some of which have major tributaries of their own
- These give the river basin a very high drainage density leading to unpredictability of river flow which is dependent on when and how quickly snow melts in surrounding mountain areas such as the Appalachians and the Rockies
- The river basin extends into 31 states of the USA, leading to a need to manage water supply to satisfy the increasing demand for water for domestic, power, industrial needs across such a huge area
- Increasing demands from farmers for irrigation water to try and feed increasing population
- Rainfall graphs for Denver, Minneapolis and Memphis indicate variable seasonal nature of rainfall across the river basin – eg fairly dry in Denver most of the year but heavy precipitation throughout the year in Memphis by which time most of the tributaries have already reached the Mississippi – leading to flooding and also run-off of water that could be stored and used in dry months
- Temperature graphs for Denver, Minneapolis and Memphis indicate variable seasonal nature of temperatures throughout the year leading to high evaporation rates in the summer but much colder conditions throughout the basin in winter

- Diagram Q3D indicates that there is a need to regulate flow of river to prevent flooding during peak discharge and to keep water level high enough for navigation in dry months

(c) Political problems for the Mississippi might include:
- difficulties between the 31 states which are represented by different political parties
- sharing allocation of water rights
- changing needs of different states including increasing populations and increasing irrigation
- increased pollution and salinity downstream affecting water quality
- shared costs of purification, flood prevention, navigation control and desalination plants
- impact of dam construction on consumers downstream
- relationship with Canada which has a small part of the Upper Mississippi basin

(d) Answers should be authentic for the chosen river basin. Candidates must refer to all 6 sections for full marks.

Answers will depend on the river basin chosen. However, for the River Nile they might include:

Social benefits:
- greater population can be sustained with increased food supply
- less disease and poor health due to better water supply and more food being available
- areas at reservoirs, eg Lake Nasser, give opportunities for tourism, eg game fishing for Nile perch and Tiger fish
- regulation of river flow greatly improves flood control on river

Social adverse consequences:
- people had to be moved off their land as valley areas were flooded eg 90,000 Nubians from the Aswan High dam/reservoir site
- loss of burial sites and other Nubian sacred areas. Destruction of Nubian nomadic pastoralist lifestyle
- increased incidence of water borne diseases such as Bilharzia due to snails in irrigation channels

Economic benefits:
- HEP attracted industries eg Aluminium smelting, fertiliser industries
- regulation of river flow improved navigation below the Aswan dam
- expansion of irrigated land led to improved farming outputs with possible surplus for sale
- improved communications with a weekly ferry from Aswan to Wadi Halfa on Lake Nasser
- initial reduction by up to half the sardine and shrimp stock off the delta but now back to pre-dam levels. Mediterranean fishery off the Nile delta has expanded due to run-off of fertilisers and sewage discharges – landings of fish are 3 times pre-dam levels

Economic adverse consequences:
- huge cost of building the dams eg Aswan cost 1 billion US $. This put Egypt into debt to Russia
- high cost of maintaining dams, power plants and irrigation channels
- 98% of the silt that used to fertilise the lower Nile is now being trapped behind the Aswan Dam
- The red-brick industry, which depended on delta mud, has been severely affected

Environmental benefits:
- Improved and more reliable scenic opportunities for tourist industry. Nile cruises etc.

- Lake Nasser provides a sanctuary for waterfowl and wading birds and has more than 32 species of fish
- Reliable seasonal water flow for plant and animal life

Environmental adverse consequences:
- water in river and on farmland becomes saline with high evaporation rates – farmers downstream have to switch to more salt-tolerant crops
- Poor irrigation techniques have led to waterlogging of soils
- change in river regime has caused the loss of many animal habitats eg the drying up of the Nile delta area may lead to inundation of sea water
- flooding of archaeological/historical sites eg UNESCO provided 40 million US $ to rescue Abu Simbel and 19 other monuments
- the water table is rising in the Nile valley, causing major erosion of foundations of ancient temples and monuments

Question 4 – Urban Change and Management

(a) Answers may include:
- Ten of the twelve largest urban areas expected to be in the Developing World
- Tokyo and New York are the only examples from the Developed Countries
- Concentration in Asia, where there will be eight out of twelve urban areas, including three in India and two China
- Majority on or near coast

Explanations could include the following:
- Historical / political / strategic factors in location of capital / primate cities
- Coastal / river locations for trade, communication and transport of raw materials etc
- Mainly areas of low-lying, flat land for ease of building
- Access to raw materials leading to proliferation of industrial growth
- Accessibility / route centres
- Higher birth rates and more rapid growth in Developing areas as opposed to stable or declining rates in Developed areas
- 'Push' and 'Pull' factors with improved levels of health, education and economy perceived to be present in the urban areas of the developing world. Large movements of migrants

(b) (i) Problems might include:
- Recreational and farmland used up by urban sprawl
- Sprawl threatens biodiversity / wildlife habitats and removes clean air lungs and open land
- Increased commuting leading to traffic congestion and increasing levels of air pollution
- Buildings and services in inner urban areas not being used or become run down or derelict
- eg housing, schools, factories and shopping areas

(ii) Solutions for example of traffic congestion may include:
- Policies to reduce cars eg car sharing, high occupancy vehicle lanes, new car charges
- congestion charges, cycle routes
- Promotion of improved public transport, including lower pricing, trams and integrated transit systems
- Park and Ride schemes
- Changing road systems eg flexi-time travel, tidal flows, coordinating traffic lights, bus lanes

(c) (i) • Inefficient urban infrastructure eg incomplete water and sewerage supplies and connections,
- leading to disease spreading
- Unemployment / underemployment with the growth of the 'grey' economy and black market

- Drugs, crime etc common and pose an increasing threat to public safety. Poor wages for unskilled jobs due to the huge supply of potential labour
- Lack of services, schools and hospitals
- Severe traffic congestion and associated high levels of air pollution – growth of 'informal' city transport (bringing benefits and drawbacks)
- Continued growth of 'shanty towns' often on unstable land such as steep hillsides where landslides are common or on marshland

(ii) Methods used to tackle the problems should be authentic and appropriate to the candidate's chosen city:
- Local authority plans to improve basic infrastructure, including provision of water / sewerage, power and roads to established 'shanties'
- Provision of hardware / utilities with 'self help' schemes eg Sao Paulo where the local population provide the skill / effort to install these ie the 'basic shell' of housing such as breeze blocks being supplied
- Building of high-rise apartment blocks in suburbs to provide high-density housing to replace the extremely high-density living in shanty areas
- Charity / Church groups offering support and advice to assist in raising levels of education and reducing levels of crime

Question 5 – European Regional Inequalities

(a) **Countries may wish to become members of the European Union for the following reasons:**
- Removing trade barriers to boost growth and create jobs.
- Tackling climate change and promoting energy security.
- Improving standards and rights for consumers.
- Fighting international crime and illegal immigration.
- Bringing peace and stability to Europe by working with its neighbours.
- Giving Europe a more powerful voice in the world.
- Securing food supplies and essential raw materials.
- Improving standards of living in the member states.

Specific EU measures to aid development include:
- European Regional Development Fund (ERDF) which provides a wide range of direct and indirect assistance to encourage firms to move to disadvantaged areas eg loans, grants, infrastructure improvements.
- European Investment Bank (EIB) provided loans for businesses setting up in disadvantaged areas.
- European Social Fund (ESF) assists with job retraining and relocating
- Cohesion Fund – aimed at states whose Gross National Income (GNI) is <90% of EU average.

(b) The three indicators given all identify a similar pattern identifying regional inequalities within the UK – East Midlands, East England, SE and SW England and London generally fare better than Wales, the West Midland and areas further north.

Candidates should use some form of comparative statements covering all three indicators to get full marks.
- Population change – Regions with the highest projected increase in population are in the south of England, both for the 5 year and 10 year projections. Much slower population increase in Scotland and NE England. Wales, NW England, Northern Ireland and West Midlands have an intermediate growth rate.
- Average House Prices – highest in London and SE England while Scotland, Northern Ireland, the 7 most northern regions in England and Wales have figures well below the UK average.

- Gross Disposable Household Income – similar to average house prices, with East Midlands and Scotland in an in-between position but Northern Ireland, NE England and Wales worst off for this indicator.

(c) The UK's regional inequalities stem from a combination of the physical differences between the higher and steeper land to the north west of the UK compared with the lower and more gently sloping land to the south and east coupled with the remoteness of the north-west compared to the proximity of the south-east to the 'core' of the EU. Candidates may justifiably stress the positive and negative aspects of different regions.
- Physical factors might mention advantages / problems such as relief, rock types, climate and water supply, soil fertility and erosion
- Human factors might mention decline in traditional heavy industries, growth areas of new lighter industries and hi-tech industries, out-migration from north and differences in accessibility related to communications and remoteness

(d) UK national government help could include:
- Regional development status
- Enterprise Zone status
- Capital allowances, training grants, assistance with labour costs
- Specific assistance to former coal mining/iron and steel areas
- Intervention of national government resulting in the relocation of major government employers or state owned firms to disadvantaged areas eg DVLA in Swansea, MOD in Glasgow
- Tesco Finance to Glasgow - £5 million Regional Selective Assistance (RSA) grant

Comment should be made on the effectiveness of the measures outlined eg the long term benefits or disadvantages of using these incentives.

Question 6 – Development and Health

(a) Social indicators could include:
- Average life expectancy at birth in years
- Infant mortality rates per 1000 live births
- Adult literacy rates (percentage)

Economic Indicators could include:
- Gross Domestic Product per capita US$
- Average Annual Income per capita US$
- GNP per capita US$
- Percentage of working population in Primary Sector

(b) Candidates should be able to refer to:
- Oil rich countries such as Saudi Arabia; well-off countries like Malaysia which can export primary products such as hardwoods, rubber, palm oil and tin.
- Poor Sahelian countries like Mali, Chad and Burkina Faso which are landlocked, lack resources, have poor quality farmland, high levels of disease.
- Newly Industrialised Countries eg South Korea, Taiwan have high GNPs due to steel making, shipbuilding, car manufacturing, clothing etc. countries with entrepreneurial skills and low labour costs.
- Large countries eg Brazil have variety of opportunities ranging from resources in Amazonia to tourism in South East Brazil around Rio.
- Tourist destinations eg Sri Lanka, Thailand, Barbados earn foreign currency and improve living standards and create new job opportunities.
- Countries which suffer natural disasters which restrict development and cause massive damage to infrastructure eg drought in Somalia, floods/cyclones in Bangladesh, hurricanes in Caribbean, earthquakes in Haiti and tsunamis in Indonesia.
- Mountainous countries eg Tibet which restrict communications and farming.
- Areas of political instability which diverts aid and resources away from areas of need eg civil war in Sudan, conflict in Afghanistan, corruption in Zimbabwe.

(c) (i) Answers will depend on the disease chosen but for malaria might include:
- Physical factors:
- Hot, wet climates such as those experienced in the tropical rainforests or monsoon areas of the world
- Temperatures of between 15C and 40C
- Areas of shade in which the mosquito can digest human blood

Human factors:
- Suitable breeding habitat for the female anopheles mosquito – areas of stagnant water such as reservoirs, ponds, irrigation channels
- Nearby settlements provide a 'blood reservoir'
- Areas of bad sanitation, poor irrigation or drainage
- Exposure of bare skin

(ii) Measures taken to combat malaria may include:
- insecticides eg DDT – however this is environmentally harmful – impacts on the food chain and is supposed to be banned as a result. In addition the mosquitoes build up a resistance to chemical insecticides through time and they become less effective
- newer insecticides such as Malathion – these are oil-based and so more expensive/difficult for developing countries to afford – also stains walls and has an unpleasant smell – so not popular
- mustard seed 'bombing' – larvae become wet and sticky and drag mosquito larvae under water drowning them
- egg-white sprayed on water – suffocates larvae by clogging up their breathing tubes (as with mustard seeds – wasteful, costly and fairly impractical)
- BTI bacteria grown in coconuts. Fermented coconuts are, after a few days, broken open and thrown into mosquito infested ponds. The larvae eat the bacteria and have their stomach linings destroyed. Cheap, environmentally friendly and 2/3 coconuts will control a typical pond for up to 45 days
- larvae eating fish eg Nile Tilapia, Muddy Loach – effective and a useful additional source of protein in people's diets
- drainage of swamps – requires much effort – not always practicable in the Tropics.

Treating those suffering from Malaria:
- drugs:
  ° Chloroquin – easy to use/cheap but mosquitoes are developing a resistance to it
  ° Larium – powerful, offers greater protection but can have harmful side effects
  ° Malarone – fairly new drug – said to be 98% effective – few side effects but very expensive
  ° Vaccine – still being developed/not yet in widespread use (eg Dr Manuel Pattaroya's in Colombia)
- education programmes:
  ° insect repellent eg Autan
  ° cover skin at dusk when mosquitoes are most ravenous
  ° sleep under treated mosquito nets – fairly cheap

- herbal remedies : Quinghaosu – extracted from plant – used as a traditional cure in China for centuries – now in pill form – easy to take – may be the long awaited breakthrough

No one solution has been found. A combination of strategies/control methods, combined with increasing public awareness/education programmes (eg WHO's 'Roll Back Malaria' – a global campaign aimed at halving the number of malaria cases by 2010) will be needed just to keep malaria in check. Some progress may be made thanks to the millions which the Bill and Melinda Gates Foundation has set aside for research into a cure.

(iii) The benefits of controlling the disease on a Developing country might include:
- Saving money on health, medicines, drugs, doctors etc.
- Reduction in national debt
- Healthier workforce and increased productivity
- Longer life expectancy and decreased infant mortality
- Money available to be spent on education and infrastructure
- More tourism/foreign investment attracted to country, leading to more employment, increased prosperity